Con

Soulways

The developing Soul–Life Phases, Thresholds and Biography

RUDOLF TREICHLER

Translated by Anna R. Meuss
and Johanna Collis

HAWTHORN PRESS

Published by Hawthorn Press
Hawthorn House, 1 Lansdown Lane, Lansdown,
Stroud, GL5 1BJ, United Kingdom
Tel. (01453) 757040   Fax. (01453) 751138
English translation © 1989 Hawthorn Press
Reprinted 1996

Translated from the German, *Die Entwicklung der Seele im Lebenslauf; Stufen, Störungen und Erkrankungen des Seelenlebens.* 2. Auflage 1982. Verlag Freies Geistesleben GmbH Stuttgart

Acknowledgements

Grateful acknowledgements for assistance with the publication of this book are made to members of *Transform*

British Library Cataloguing in Publication Data
Treichler, Rudolf
    Soulways
    1. Psychotherapy
    I. Title   II. Series
    616.89 14

ISBN 1 869 890 13 2

Cover design by Patrick Roe., Glevum Graphics
Cover picture by Ivon Oates

Printed in the United Kingdom

Preface

There is growing interest in biography and personal development. Many people are coming to realize, or at least have the feeling, that it will be easier to cope with life's problems and crisis situations if one goes to the roots and sees how they have come about. Planning and decision-making are more effective if the background and origin of a situation are considered first, before a link is established between past and future that will also relate to present difficulties.

The study of any biography will show it to be a process of development. The purpose of this book is to consider the development of soul life in relation to biography. This will be seen to follow certain laws, knowledge of which makes it easier to cope with life's crises and upheavals. It will be shown that–in conjunction with other, contributory factors–upheavals and upsets may eventually lead to mental and emotional illness. The vast field of psychiatry opens up at this point, and it is also the aim of this book to show how a more comprehensive view may be gained in this field.

There are good and practical reasons for this, for disorders of mental and emotional development and mental illness are constantly on the increase. At the 22nd International General Medical Congress held at Marburg, West Germany, in 1977, it was said that approximately every third patient seen by a general practitioner suffers from mental illness. If one then considers that there are many other individuals with less serious mental and emotional problems who do not go to see a doctor or psychotherapist, it is obvious that every human being who has a feeling for others is now called upon to give what help he can. Effective help can however only be given if there is at least some understanding of the real nature of the developmental disorder or illness. A further purpose of this book is to convey an idea of this.

The book is intended for physicians, teachers in remedial education, school teachers, priests, counsellors and psychologists, and in the

final instance for anyone who takes a real interest in mental and emotional development and is encountering problems in this area-
problems affecting himself or others. Some basic knowledge of the
anthroposophical approach to the study of man will be a help to the
reader. For anyone who does not have this knowledge, the works
of other authors, predominantly Rudolf Steiner, are quoted in the
text. These works also served the author as source material. The content of this book will only prove fruitful to the reader, and can only
be judged, if that basic knowledge is available. Systematic knowledge
of psychology or psychiatry is not required, but in this respect, too,
frequent reference is made to works that will help the reader to learn
more about the subject.

    The author is also indebted to the work of other authors who have
provided pointers, additional information and further confirmation.
The book includes ideas and the experience of many colleagues and
people who have attended courses, and with whom the author has
been working, in some cases for decades, at the Friedrich Husemann
Clinic in West Germany, in psychiatry courses at the Goetheanum
in Switzerland, and many medical conferences, seminars and groups.
The author feels particularly indebted to Friedrich Husemann, the
first anthroposophical psychiatrist and founder and head of the clinic
that was later to bear his name. The impulse to write this book arose
out of the need of mentally ill patients and those with developmental disorders whom the author sought to help. His gratitude extends
to them as well. The greatest debt of all is owed to Rudolf Steiner,
the creator of anthroposophy, the discipline on which this effort to
extend the horizons of psychology and psychiatry is based.

Introduction:
Soulways and Biography Work

Soulways charts the unfolding of the human soul through the emerging life phases, transitions, thresholds and pathways–the seasons of human life.

The title may sound unusual to those brought up in a scientific tradition which reduces psychology–the study of the soul–to behaviourism. A premise of behaviourism is that only observable behaviour can be studied systematically. Any 'inwardness', 'awareness' or 'soul' is passed over–we are merely 'behaving beings'. For example, the late John Davy, once interviewed a leading behaviourist, B. F. Skinner, author or *Walden II* and *Beyond Freedom and Dignity*. Skinner commented at the end of the interview, words to the effect that, ' . . . we have not been conversing, we have merely been behaving . . . '.

However, *Soulways* follows such books as Gail Sheehy's *Passages*, Daniel Levinson's *The Seasons of a Man's Life* and Bernard Lievegoed's *Phases*–in mapping adult development from a psychological and spiritual perspective. Of course, as a psychiatrist, Rudolf Treichler recognises the importance of the somatic, biological influences on human development and health.

The translation of *Soulways* was made as a response to the needs of a network of people doing what is becoming known as 'Biography Work'. Various counsellors, trainers, management and personal development groups such as Transform, felt that *Soulways* would help both deepen their work and help readers come to a deeper understanding of their own biographies, of 'what they have written in their lives.'

Biography work arose in the 1970's in Britain when, as a result of management and organisational development programmes, participants asked 'for more', to help them face choices in their lives and work.

We all reflect on our lives from time to time, even though we are told of the dangers of 'navel watching' or 'paralysis of analysis'. Biography work is for those who feel that an exploration of their lives can lead to insights, to learning, to new ways of being and to more conscious choices.

The main benefit of working biographically is that it enables people to take small steps in their development. For example, one may be at a crossroads and wish to consider options. There may have been setbacks, conflicting values, outmoded roles, illnesses confusing relationships, emotional limits, inner blocks, the experience of recurring life patterns or questions of personal growth.

People facing such situations might benefit from doing biography work. Usually, this work is carried out by individuals in small groups, and by individual reflection. There may be workshops, or weekly meetings, and counsellors may use biography methods on a one to one basis.

The aim is to enable people to explore their own questions in relation to their biography. Practical exercises and artistic exercises support individual reflection and group work. There may also be talks on life phases, the meaning of biography, barriers and breakthroughs in development, the roles we adopt and are given, viewing our lives through our temperament, creativity–becoming ourselves and preparing for the future.

Biography work has many links with other areas, such as developmental counselling (as opposed to remedial or re-habilitative counselling), career and life planning–and is used frequently at the start of management self-development groups to identify issues. However, it is becoming a distinct strategy for individual development as the pressures of modern life increasingly challenge us to take up the threads of self awareness and human responsibility.

Biography methods are also used to help those considering setting up their own business to do so in the light of their lives as a whole. Such methods have also been used to assist individuals face redundancy, retirement and organisational development. Men's and women's groups working on gender questions have found the biographical approach to be helpful.

An understanding of biography can become a threshold to deepening relationships. When we know the biography of another person, and share our own, a real meeting can take place. When one meets another person in this way, everyday surface judgements are replaced by understanding and love. An awareness of the issues at the heart of *Soulways* and of biography work, may lead to the recognition of each person's biography as a 'project' with unique challenges, potentials and qualities. Instead of life being fired at one from point blank range, and the 'I wasn't ready for this' feeling, one can take up 'biography tasks'. Then, according to an ancient Chinese philosopher:–

> The human being becomes master of the changes of nature if s/he recognises the sequence and adjusts to the succession of the seasons. In this way come order and clarity into the apparently chaotic changes of the times, and people can adapt themselves beforehand to the demands of the different phases *

Martin Large

*From Esther Harding, *The Way of All Women*, Longmans 1936. I am indebted to Jenny Daisley's handout, *Working with Biography* and her Transform brochure, *Biography work*, for background material for this introduction.

Chapter One
Human Biography–Fundamentals

A human biography is not an event that merely takes a predetermined course. With the aid of memory we can perceive, first of all, that our own biography has a time form, that it has grown into a differentiated organism and is going through a process of development. We will find the same to be true for the biographies of others. Every biography shows us a human being who is going through a lifetime of development. Before a biography can be put down on paper, the individual has already 'written' it, inscribed it in the world, by living it. A biographer is merely copying from the 'original'.[1]

Plants and animals also have life; but in their case the biography is not 'written' by the individual plant or animal. The life of a plant is entirely determined by the cosmos. The plant takes its orientation above all from the sun, and in the process of assimilation the light of the sun causes the plant form to condense from the air.

Animals emancipate from the outside world to some extent; in their case the life of the cosmos is internalized as the organism closes in upon itself. Because of this, animals are able to react to the world; their kind of body enables a soul to have sensations and to act. An animal is not open to the light of the sun in the way a plant is, and therefore the light of conscious awareness arises in its soul.

Individual animals are however always bound up with the species of which they are members. An individual lion is species-specific and not individual in his reactions. The stirrings of an animal soul do not arise from a centre within, but stream through the animal from the species; they are all-present in the animal's soul life, providing the impulses and inescapably determining its life.

In the human being we finally find a centre that provides the basis for individual life. In addition to conscious awareness in the soul there is awareness of self, of being an ego; the human soul has an innermost core; everything it experiences relates to this, and all its responses originate from it. Animals do not have an ego and are determined by the 'group ego' of their species. Every human individual on the other hand has his own ego.[2] The light of conscious awareness has thus gained a focus, a centre, that initially is like a spark of light in the soul. The 'small spark of divinity' of Meister Eckhart that glows in every human soul becomes the inner sun in Goethe's words:

> And then that conscience of your own
> Shall be the Sun within your moral day.[3]

The words 'small spark of divinity' indicate that because of the ego, a divine principle is able to incarnate. If we live in the light of our inner sun, we establish harmony between our own life and the divine and spiritual world. Yet that inner light of ours is merely a spark of the divine light. 'The ego receives the rays of the divine light that shines forth as the eternal light in the human being.' Because of the ego, the life of the spirit takes individual form in the human soul.[4]

The essential nature of the human being reveals itself in every biography. A plant has a living body; animals can be seen to have an ensouled living body. In human beings the ego is also present, and this is a manifestation of the spirit. Only beings endowed with body, soul and spirit can create a biography in the ensouled physical body and inscribe their biography in the world.

Biography and Rhythm

Like all growing things, the time form we call a 'biography' is given an infrastructure by different rhythms. The most important rhythm in human life is the seven-year rhythm.

We find that life goes through a fundamental change about every seven years. This is most obvious when the second dentition develops at around the seventh year and when puberty starts in around the fourteenth year. Careful observation will however show that such changes occur throughout life. To say 'about every seven years' does not suggest a lack of scientific accuracy but has to do with the nature of rhythms. According to Klages, a 'beat' results in repetition at regular intervals whereas 'rhythm' is repetition at similar time intervals.[5] Fluctuations are therefore part of the seven-year rhythm in human life. Key elements in this are anticipation and delays, and these are due to the evolution of humanity on the one hand and to the nature of the individual on the other. Earlier onset of second dentition and puberty is a phenomenon relating to the whole of humanity; time variations in developmental periods in later life have more to do with the nature and destiny of the individual.

In one of his fundamental lectures on mental and emotional development that have been published under the title *Metamorphoses of the Soul*, Steiner said that when he spoke of the seven-year periods this always meant approximate figures.[6] This should be remembered, but it is nevertheless right and indeed fruitful to consider the seven-year rhythm the basic rhythm in human life. Hoerner has published comprehensive studies according to which the figure seven plays a dominant role in the whole cosmos. In the seven days of the week this provides for a 'rhythm of the soul' where each weekday relates to one of the seven planets that also relate to the soul.[7] The significance of the figure seven goes beyond the planets and the life of the soul, however, for it also determines a number of biological rhythms.[8] On the other hand the rhythm of the physical body, with its annual birthday, is that of the year with its twelve months.[9] The seven-rhythm of life and of the soul combines with the annual rhythm of the physical body to give a seven-year rhythm, and this establishes the link between the essential nature of the human being and his existence in the physical world.

The Seven-year Rhythm and Development

The essential nature of a human being does not present itself all at once but only emerges gradually, thanks to the element of time. This is what we call development, though it should not be considered to consist in the 'unfolding' of something that is already in existence. The human being has to come to terms with the phenomenal world, and further development, 'enhancement' as Goethe called it, results from the encounter with, and coping with, contrasting elements or polarities.

Development does not proceed at an even pace, however. The seven-year rhythm that gives structure to human development leads to stages in life where new potentials arise and the old can be transformed into something new.[10] Buehler was right to stress that rhythms are not so much a repetition of similar elements at similar time intervals, but rather 'a similar element coming up again in a living way'.[11] When something comes up anew, enhancement is possible and at the same time the essential nature emerges more fully than before. Rhythmical alternation of polar opposites creates a balance between them, and at the same time space is made so that the essential nature can gradually emerge. The periods of life may be considered in their biological aspect. It will then be found that 'the most basic processes in life are periodic, as modern biology has shown'. Every seven years, 'the wave of an individual life comes to a momentary halt'.[12] Those are the nodal points in life that are potential crisis points. We need to know them if we wish to assess a biography and to be able to help ourselves and others in times of crisis. When the crisis has passed it will usually be found that something new has arisen from that nodal point, just as something new arises from a node in plant life.

That is the biological aspect. Steiner added to this from the point of view of spiritual science by calling the process that occurs every seven years a 'birth'.[13] This is an image that enables us to grasp the development of essential human nature in its fullness. As in the case of a physical birth, the births that happen at seven-year intervals are more than the yielding of new fruit from a womb. New essential aspects reveal themselves that cannot be explained as having arisen from the old. From the point of view of spiritual science, a child is not merely the product of its parents, nor does it merely emerge from its mother's womb, for it also arrives. It descends from the womb of a spiritual world where it existed as an independent entity before birth, and incarnates in the earthly body that is the fruit of the maternal womb.

Subsequent births also see something essential incarnating in a physical body and in the life of the developing individual. Life shows that apart from something new emerging from what has gone before, something completely new also enters into life. It is quite right to speak of development taking a new turn so that something can come to realization in the individual life and in the world, something that depending on the dimension of that life is important, and indeed necessary, for a small or possibly even a large part of the world.

This new development is far from alien, however, it is not something coming in from outside; it has totally united itself with the essential nature of the individual. As the essence of the individual gradually comes to realization, the new element is presented to the phenomenal world as part and parcel of that individual's true nature. If the individual is a prominent person, such 'self realization' may go along with and merge into the realization of trends and goals in the evolution of humanity.

The seven-year rhythm, too, does not come from outside but relates to an inherent order in human nature. We will be considering seven-year periods when development concerns mainly the body, and others when the emphasis is more on the soul or the spirit. Every seven-year period coming under these three categories in turn relates to a further integral system within human nature. Steiner called these 'essential or constituent principles'.[13] He was concerned to make it clear that he was not speaking of a division or classification, but of an organic system the individual elements of which do not simply exist side by side. Body, soul and spirit are three essential principles or constituents. In the same way, the powers of individual constituents mingle and interact at finer levels of organization, with one principle always providing the keynote. Our limbs are bodily members; equally, every other constituent principle of the human being has its own mode of taking hold of the world and being creative in it.

The births that occur in human biography are concrete events, with one essential principle incarnating at every birth. During embryonic life the ego was involved in giving form to the growing physical body; during subsequent birth events it always also plays a role in the development of the new 'fruit'. It will be seen that particular importance attaches to this when we consider soul development.

Rhythm and Developmental Psychology

Opinions differ when the question of rhythm arises. Some time after Steiner, Guardini devoted himself specifically to the nature of rhythmical phases in life. According to him, these phases are 'configurations of life that cannot be derived one from the other'.[14] Something new enters into the life organism with a different phase, and that is why one phase will not provide the explanation for another.

Many scientists who have realized that there are rhythms in human life have come across the seven-year rhythm. Even the Greeks knew of it; they divided life into heptomades, ten seven-year phases.[15]

After Boehme and Paracelsus, it was not until the beginning of the twentieth century that modern science developed an interest in the seven-year rhythm (Swoboda 1904; later Klages and Hellpach).[16] Moers referred to six life stages, essentially in a seven-year rhythm.[17] Kuenkel described fourteen-year periods in life.[18] Most recently Sheehy has written a bestseller in which she states, out of considerable life experience and sound feeling, that our inner life system consistently develops its own rhythm.[19] She asks if there is any truth in the popular notion of a seven-year itch in adult life.[20] And she speaks of considerable statistical evidence of the existence of a seven-year rhythm,[21], though she does not go into detail.

Other scientists, like Buehler for instance,[22] may not have discovered a seven-year rhythm, but they certainly noted distinctive stages in life. Oerter on the other hand considers development a 'complex web of cause-and-effect relations' that do not relate to any particular age level.[23] Oerter refers 'only to experimentally or quantitatively (i.e. statistically) demonstrable facts.'[24] Other authors take the same line; like Oerter they make a psychological study of human behaviour - and do not go beyond this ('behavioural psychology'). The impression is that because of its devotion to detail this particular school is unable to see a particular life as a whole and perceive its rhythms. (For a more detailed discussion and classification-of the different schools concerned with biographical studies, see Lievegoed's book *Phases*.[15])

The figure seven is merely an approximation; it does not refer to fixed points in life, and there are all kinds of variations in the seven-year rhythm. It is questionable whether such a rhythm can in fact be determined by the algebraic methods used in statistics. The seven-year periods are approximate intervals reflecting an order in the spiritual dimension; because of this, the seven-year rhythm can always be perceived as a fundamental principle in human life and one can work with it. Human life itself has its origin in the spirit–as has already been mentioned. This life comes from the spirit, is an incarnation of spirit and leads to an ascent to the spirit, as will be shown in the following chapters. Life itself demands that its course shall be subject-to a spiritual ordering principle. When the study of biography consists-in relating that ordering principle to a life, the biography becomes a work of art.[25]

Chapter Two
Developmental Phases of Soul Life

The soul's earth life begins with the first breath. At this point the soul enters into the life of an independent body. It has however already been involved in the development of that body, together with the ego. The soul now begins to be active in this life, for it exists to serve it. Because of this the soul does not initially reveal itself as clearly as it does later on, for the time has not yet come for an independent soul life to evolve.

We can feel how small children are cocooned in soul and have a soul element all-present within them. The small body is like a seed from which the soul will flower one day. For the moment the soul mainly works in and on the body itself and can therefore be experienced in the sphere of the body; one gets the feeling, however, that the soul is still veiled. During embryonic life, the body was held within the mother's womb. The soul that is active in the body–Steiner called this soul aspect the 'astral body' in the narrower sense –is held within an enveloping astral form from which it will be 'born' to live a more independent life once puberty sets in.[11] This enveloping maternal element provides more than just protection. Previously the mother's womb enveloped the child and mediated between it and the outside world. Now the enveloping astral form mediates between the cosmos, the outside world, and the soul's activity in the body; this soul shimmering in the body sometimes flashes or shines out at us.

The first two seven-year periods may thus be considered to be preliminary stages in soul development. We shall be able to deal with them fairly briefly, for they have been widely discussed by a number of authors in the field of education.

The First Seven-Year Period

The first seven-year period sees the development of the physical body, for this was far from 'finished' at birth. Many essential structures and organs only begin to develop, or develop further, after birth. The white matter of the brain, which provides the physical basis for the faculty of thought, essentially only develops after birth (maturation of medulla). It is only in the eighth year of life that maturation has 'progressed to a point where there are no longer any significant anatomical differences from the adult nervous system.'[1] The buds of the permanent teeth are in essence established around the seventh year of life and the permanent teeth will gradually erupt to replace the milk teeth.

The permanent teeth are the last physical organs to develop, and this marks the end of a creative activity in the body that will never again be repeated in just that form. Any further anabolic organic activity serves the purpose of counteracting the continuous catabolism that takes place throughout life; it will be re-creation not new creation. The child has now formed its own body out of the 'prototype body' that it received through inheritance and used as a model. In about the seventh year the process reaches its conclusion when 'a human being's own teeth replace the inherited teeth.'[2]

The biographical significance of childhood diseases begins to emerge. A childhood disease can help a child to penetrate the inherited body with its own powers.[3] The faces and eyes of children recovering from measles reveal more of their individual souls that are working on their bodies than could be seen prior to the illness. The children are more awake and also more clearly individuals in the way they now look on the world. This, however, only marks the beginning of a process in which the soul (with the ego) enters more and more into the physical body, a process that will continue until adulthood is reached. The need to come to terms with heredity also continues. The inherited model does have a role to play in later life. Weaker individuals will continue to follow this model after the first seven years, stronger ones will be less inclined to do so. According to Steiner, inherited elements in the physical body also influence the astral body in its activity within the physical body, i.e. also the form that 'soul qualities' take.[4] This points to hereditary traits as a factor in the development of mental illness in adult life. (See the chapter on Psychiatric Conditions in relation to Biography, p. 150 ff.)

The seven-year period of the physical body is also of fundamental significance for the development of the soul once it has grown independent, for later on the body will be the instrument of the soul and support soul life.[5] The first seven years are vital in determining the form and structure of this instrument. Physical growth will continue beyond them, but 'throughout subsequent life, such growth is determined by forms that have developed up to the above-mentioned point in time.'[6] Depending on the way these forms develop, the soul will later find it easier or more difficult to fashion its individual life.

The child's essential nature is not the only factor in the way the body develops; the environment also plays a role. In fact, the responsibility of the environment is greatest during the first seven years, when the child has left both the cosmic and the maternal womb and is asking the earthly world to take care of it. The spiritual world continues to guide the child, but in its creative activities in the earthly world the child is now more than ever dependent on the environment. Its physical body is, after all, part of this.

The constructive (or destructive) factors in the environment that influence the child's body join forces with the soul which acts as the mediator between body and outside world. The senses enable the soul to perceive these factors, and everything that is perceived by the soul has an immediate formative influence on the body. Apart from anything else, this shows that a child's soul lives mainly in the head which is the centre for sensory activity and at the same time also the source of the 'process that shapes the human form'.[7] This process is made possible by the nervous system, which again has its centre in the head. Transmitting sensory impressions from without to within, it provides the basis on the one hand for conscious ideas to be formed and on the other for formative processes that remain unconscious. In children, the emphasis is initially on the formative function of the nervous system; the paralyzing, image-forming activity that leads to ideation will only develop gradually at a later stage and then become dominant.

During the first seven years of their life children are 'imitators'.[7] One often sees this. It is only rarely, however, that such imitative behaviour is conscious. The subconscious imitation practised by children relates to more than just emotional and physical reactions: the sensory impression underlying those imitations leads to a kind of organic imitation being made in shaping the physical body. The senses enable the soul to move in tune with everything that happens in the environment, and this movement, of which there is only little conscious awareness, reaches the physical body. The child's form is 'filled with soul, filled with spirit'[7] from the head. Unlike the head of an adult, it is very much alive, and it is from the child's head that the soul relates to the life of the body and feels its way into the body and into the world.

The fact that the spirit penetrates the body from the head draws our attention to the child's ego. This works both in and together with the soul. Initially working from the head, it has made the child stand erect and walk upright. During the first seven years the upright posture and gait is governed and supported from the head, just as a string puppet is controlled by strings.[8] In around the third year, when the child begins to call itself 'I', the light of self awareness dawns in the head. From this point onwards, everything the child takes in through the senses is also taken into its self awareness, and the ego is then able to perceive it again inwardly in the process of memory. Self awareness is slowly growing and beginning to put its imprint on all the ways in which the child's soul comes to expression.

The Second Seven-Year Period

The end of the first seven-year period also means the end of creative organic development in the physical body. It means the release of creative powers. These are part of the organism of creative powers that has been mediating the transmission of form-giving impulses of soul and ego via the nervous system to the physical body. The greater whole, the 'entelechy' that according to Driesch is behind the differentiated forms in the physical body and integrates them so that they are an organism,[9] assumes concrete significance in the essential principle that Rudolf Steiner called the 'life body' or 'ether body'. The word 'body' signifies that the creative powers–and in the case of the astral body powers of soul–have assumed a specific 'configuration', though this is not perceptible to the senses,[10] and are indeed an organism. The term 'ether' points to the origin of the creative powers in the 'ether' that surrounds the earth. The term 'astral body' refers to a sphere that cannot be perceived by the senses and of which the stars are to some extent representative. The soul descends from this sphere at birth and returns to it when we die.

When creative powers are released from involvement in physical activity this signifies that the organism of creative powers is coming free of the physical body and a second birth occurs in the biography. The ether body is born in around the seventh year, and the seven-year 'period of the ether body' then begins. It is in the region of the head that this birth results in the highest degree of separation from the physical. The buds of the second teeth are now fully developed, the brain achieves full 'maturity', and life is definitively withdrawing as conscious awareness develops. The child has come more awake, is able to learn and therefore to go to school.

The ether body acts as mediator between soul and body, and it is due to this fact that the creative powers of the physical body are now becoming available to the soul. 'Powers of organization and growth' become 'powers of thought'. Steiner even put it like this: The 'human being's powers of thought are a subtler form of powers of form and growth.'[11] It is therefore reasonable to consider how the formative principle of what initially were creative powers giving rise to the physical body, now takes effect in the sphere of soul life. Here further transformations take place and continue a process that started in about the seventh year of life.

Metamorphosis of Powers of Growth into Powers of Thought
The physical body is created from cells, and in the process the 'obstinately independent life' of the cells has to be partly overcome if the organism and its organs are to have their own organization.[12] It is through the mediation of the ether body that the 'process of creating the human form' overcomes the cell principle. On the other hand the ether body also lets cells grow; it needs them in order that it may be 'taken up' by the physical body. 'No living being ... can have an ether body unless it also has cells.'[13] The process that gives rise to physical cells is a continuous one, with the actual cells always representing 'end states'.[14] The ether body needs them for its own support and also integrates them to produce physical organs.

The life of the soul is also in constant flux; it yields end products that may combine to form 'organs': These are concepts, the final outcome of judgement and evaluation. Judgement that 'the rose is red' develops into the concept of 'the red rose'.[15] The process of judgement or discernment that occurs in the soul as it experiences life has greater substantiality than the idea derived from it; we have greater awareness of ideas, for they live on in the soul and can be brought to mind again. Just as a body is made up of cells, so the life of the soul is made up of concepts. Like cells, these have a certain 'life of their own' and also help us to understand anything new. Steiner spoke of 'a body of' previously formed 'concepts that makes comprehension possible'.[16] Unlike cell structures, which occupy space, concepts are not constantly present but are recalled again and again. Subject to this limitation they show the highest degree of form and substance in soul life and are comparable to the cell principle at the physical level. Cells are combined in organs, and in nutrition and respiration these assimilate contents derived from the physical world outside. Concepts combine in the soul to form organs that serve to assimilate contents derived from the world of soul and spirit.

Individual concepts first of all combine to form 'conceptual complexes'. Things we have learned and that have become conceptual complexes 'go to sleep' when we forget and 'wake up' again when we remember them.[17] In the process of learning they become the organs that enable us not only to recognize world contents again but also to assimilate, i.e. come to understand, them.

When we remember things, more may come to mind than we had originally forgotten; remembering something means to grasp more of it than we did when we first perceived it. For example we may have developed a number of concepts relating to the shape and the growth sequence of a plant. Those separate concepts then form an organ in us that enables us to understand the plant as a whole when we call it to mind again. Having forgotten, we even find that once a conceptual complex is called back to mind the old organ has undergone a transformation. A new, more developed organ has taken its place, and this enables us to gain new insights into the nature of the object. These are the organs we use in our thinking, and they are much more alive and mobile than physical organs—as one would expect, considering their etheric nature. They are subject to constant renewal. As Goethe said, 'Every new object that we look at carefully, brings out a new organ in us.'[18]

To sum up: The creative powers that initially induced cells to grow and integrated them into physical organs are transformed and then induce concepts to grow and finally develop into organs of soul and spirit. In the metamorphosis of those creative powers the ether body acted as a mediator between the sphere of the body and that of the soul. The developmental stage reached by the seventh year may be called 'conceptual maturity' (Fucke).[19]

The Life of Feelings

Separation between ether body and physical body is greatest in the region of the head. It is less marked in the middle and lower regions of the organism. This is evident from the fact that the latter retain greater physical vitality than the head does. The birth of the ether body is only partial in them, and soul life does not come to the same conscious expression; it expresses itself more at the dreamlike or sleeping level of consciousness that pertains to the life of will and feelings. This leads to the 'development', indeed the 'transformation', of adopted habits, inclinations and temperaments.[20]

Quoting Gruhle, Koenig said that temperaments belonged to a 'realm that lay between body and soul', i.e. the realm of the ether body. 'In earliest childhood the true temperament' is overlaid with a 'temperament of another colour' that will gradually disappear. 'Towards the end of the second decade the final temperament emerges quite clearly';[21] it obviously needs the growing independence of soul that develops between fourteen and twenty-one to come fully into its own. 'Even in the lower classes at school, children's temperaments often play a crucial role.'[22] They provide the basis for soul life. On the one hand they determine the physical constitution, on the other the life of the soul, its pace, the way its processes run.

Between seven and fourteen years of age they above all provide the basis for the emotional life that begins to dominate soul life, taking over from the perceptive life of the first seven years. The children are now 'completely given up to soul experience'; they alternate between 'sympathy and antipathy, pleasure and pain, fear and courage'.[23] Some will become hypersensitive; Steiner characterized this as a constant 'soreness of soul'.[24] Others will grow less sensitive and may even become dull and apathetic.[23]

During the second seven-year period we experience soul and ego in the middle region of the organism, in the rhythmical system, for they have descended to this region from the head. Children live much more in their rhythmical processes at this stage. 'Before, they were engaged in modelling their own bodies (individual form); now they begin to be musicians, unconscious musicians who work in a more inward way.'[25] The soul goes inward in this musical activity of body and soul. In the rhythmical sphere the ether body is bound up more closely with the organic principle than it is in the head, but from the seventh year onwards it nevertheless rises to meet the soul and enables it to relate to the physical breathing process. 'Maturity of breathing' is achieved–a concept presented by Steiner that was taken up more fully by H. Mueller-Wiedemann in his fundamental work on the second seven-year period. Maturity of breathing makes it possible for a new way of thinking to develop.[26]

The lung now becomes the chief instrument of the feeling soul. With every inspiration the astral body, which primarily belongs to the airy element, unites more profoundly with the physical body, in which the creative ether body is active. With every expiration it separates to some extent and gives itself up to the environment. In the process of respiration the feeling soul swings to and fro between body and world, between lung and world, and through the body gains awareness of the feelings it experiences in relation to the world. Each time a feeling or emotion develops 'the breathing rhythm is modified ... and a particular feeling arises in the soul.'[27] That is how the many different feelings and nuances of feeling tinged with sympathy or antipathy arise in the encounter with the world. One-sided development of this lung-based soul life leads to extreme sensitivity in relation to the world, and to a degree this is normal in children who are in their second seven-year period.[28]

This aspect of soul life is not yet as free as it will be in later life. One is still aware of a veil covering the soul life at this stage, though this is thinner now than it was during the first seven years. This is the maternal cocoon that has already been mentioned; it is intimately bound up with the ether body as well as the physical body. In the rhythmical breathing process, the astral body is again and again entering into a sphere of intense etheric activity and taking some of it along as it moves into its more independent phase. This is why compared to the more conscious sensitivity of adults, that of children is more of an unconscious tendency or habit. It is the reason why Steiner used a relatively physical term, 'soreness of soul', in describing excessive sensitivity in children.[24]

Authority and Egoity
As the soul begins to enter into the inner body it also grows more inward itself. The principle that governs development and training is no longer one of direct imitation, but an inner desire to emulate. Authority wielded with love is what children will follow, and this can also have a physical effect on the middle organism, helping the soul element to enter into that region. The emulative response arises from the feeling of love that has awoken in the child; here we progress

from the lung to the heart as the organ for the power of love. At the same time an effort is made from the centre of the body and from the germinal centre of the soul, from the feeling ego, to find an initial balance between the one-sided devotion of sympathy and the one-sided withdrawal of antipathy.

Rudolf Steiner considered authority wielded with love a most important principle in education. A story that is often told is how each time he visited the Waldorf School in Stuttgart he would ask the assembled boys and girls if they loved their teachers. The author was one of those boys. Recalling the experience in later life he grew aware of the difference in response between the Yes the Lower School would give and the Yes that was the response of the Upper School. The boys and girls who were between the ages of seven and fourteen had no reservations as they gave their reply, but some of the older pupils had grown critical and did have reservations where one teacher or another was concerned. The question could also be experienced as a challenge to establish a new, more positive relationship to a teacher, and a challenge to form one's own opinion concerning the educational value of Rudolf Steiner's question.

In the third year of the first seven-year period, development was given a special note by the ego aspect. During the second seven-year period, the ego again shows itself to be a constituent principle in its own right. In the third year of life the ego principle shone out in the life of the head. At the beginning of the second seven-year period it is still completely embedded in the rhythmical system, in the life of feeling. In the ninth or tenth year it specifically connects with the lower or metabolic pole in the organism.[29] A number of physical changes occur that make this quite clear. The power of uprightness that so far has been acting from the head now acts from below upwards: the child comes upright from below upwards. Body heat, another sphere where the ego comes to expression, now arises through metabolism.[8] The fact that the ego is now acting more strongly out of metabolism, and particularly through the liver, is evident from the change in blood sugar level. Blood sugar is the vehicle for the ego that lives in the body; from the ninth year onwards blood sugar levels rise again.[30]

As for the soul, we note that children experience a greater degree of separation from the world at this stage; they confront the world more than they did before; they may therefore become critical, with their will driving them into opposition. On the other hand everything may subside into a dull apathy. The roots of such an apathy developing during the second seven-year period must be sought here. Overall, an experience of isolation is typical, and this is often bound up with sadness. Authority wielded with love by teachers and parents, addressing itself to the heart, is vitally important when children are in danger of drowning in the depth of their organism. It is through the heart that new links are forged with the world and the ego is offered fresh support, though this can only be taken up if the inner impulse is there.

There is another important difference in the way the ego comes in now as compared to the way it did in the third year of life. Then it shone out into the future; in the ninth year the future is anticipated. The ego is preparing the way for the incarnating astral body which will only reach its goal in the lower human being–i.e. puberty–by the end of the second seven-year period.

Here the ego is beginning to take hold of the future in its will impulses and thus preparing for the life of the will in the third seven-year period. This, however, has its consequences for emotional life. A new, denser feeling of egoity arises at this point; isolation, and with it sadness, is experienced mainly in the emotional sphere. It can be overcome through the feeling of love. The physical reflection of this is the fact that around the ninth or tenth year the heart shows 'a sudden increase in output and also in size'.[30] Finally this is also the time when children become physically aware of their hearts.[26]

The Third Seven-Year Period

The first two seven-year periods may be called preliminary stages in soul development. We shall see that the third period represents the initiation of, or overture to, soul development. Overtures often contain the main themes of the opera that is to follow; in the third seven-year period we are able to discern the soul elements that will play a major role in subsequent periods.

The period from 14 to 21 years is the period of the astral body, and the astral body comes to birth at puberty. A new form of soul life emerges from the cocoon; it has more of a personal note, though there still is no individual personality, no ego, at the core. The ego is still entirely caught up in the soul, but now lives in it in a new and different way. The early stirrings of the will in the ninth or tenth year now become will impulses in which the ego is struggling to come to terms with the emergent powers of the new soul life. The end of the process will be a responsible individual with a role to play in the world.

To begin with, the ego experiences the dramatic problems arising from the polarities that develop in the astral body after its birth. It is in about the twelfth year–when children begin to grow taller again in what is known as the 'second change of form'–that they progress to abstract conceptual thinking.[24] The head has lost much of its vitality by now and this is in contrast to the lower pole in the organism, where new life is beginning. Pubescent young people tend on the one hand to be critical of the world around them, and this comes to expression in abstractions. On the other hand they react emotionally and with passion from the lower pole of their nature, sometimes merely out of impulse. They may be disinterested, lukewarm or apathetic, not wanting to stir a muscle; or they may give way to powerful impulses to be active and sometimes even great rushes of movement. They may withdraw to their rooms, deliberately recreating the isolation experienced in their ninth year, or they may rush into the world, sometimes becoming downright rolling stones.

A relationship to the world based on inner feelings that had its ups and downs but nevertheless continued as a relationship, has now given way to an attitude that may be rational or emotional, or a combination of both. The alternation between sympathy and antipathy of the second seven-year period has now become an impulsive accentuation of the lower or upper pole of human nature, and this may come to expression in instinctual inclination or disinclination, emotional dissolution or rational rigidity. Jung[29] spoke of a definite 'falling out with oneself' in connection with puberty, a 'dualistic phase.' He even spoke of a 'birth of the soul', though he did not connect this with physical processes the way anthroposophical science does.

This inner dualism has its external counterpart in the separation from the parents that is now taking place; this goes hand in hand with separation of the astral body from its maternal cocoon and from the physical body. It is a normal transitional stage for many young people in puberty to fall out with their parents, and if that is all it is, nothing needs to be done about it.

The Two Sexes
A further polarity arises due to the difference in development that

puberty brings for the two sexes. Boys develop and experience extremes to a greater degree; girls sustain more of their middle system and its life of feeling. With the emphasis on the middle–also anatomically–and a distinct emphasis on feelings, girls have the tendency to live and make their judgements on the basis of inner images.[31] They are more inclined to daydream than boys,[31] and that has also been a characteristic of the second seven-year period. When a young man yearns after a girl, he is also yearning after a part of his own inner nature that has been lost, his own middle from which it is possible to move either up or down.

Men are more strongly influenced by the earth; women, being less deeply incarnated in their bodies and into the earth, are more influenced by heaven, and this is through their middle region.[32] In the encounter with a woman a man may therefore experience echoes of a relationship to the cosmos that existed in the childhood of man and humanity. This comes to expression in many fairy tales, legends and poetic works, examples being the figure of Brunhild in the Siegfried legend and Margaret in the story of Faust. (Only brief reference can be made to the problem of female characteristics in men and male characteristics in women; the solution lies in the realm of the constituent principles. For this and other problems relating to the two sexes, see Leber's book.)[33]

Puberty leads to 'maturity for life on earth'.[34] This occurs in both sexes, though more strongly in the male, and establishes a new relationship to the earth. Sexual maturity is one aspect of this, with young men and women now able to unite their bodies to reproduce their own kind on earth. The creative powers that were active in the individual physical body until the seventh year can now be creative in generating another physical body. This however involves not only the ether body as the mediator of life, but also the desire-filled astral body, which has followed the ego into the metabolic region.

In the male organism the astral body has become completely immersed in this region, and indeed has gone through and beyond it. This is evident from the position of the male gonads, for these are outside the abdominal cavity and their seed leaves the body once puberty has been reached. The soul equivalent of this is the nature of male desire, for this represents the active approach to the partner. In the female organism the astral body acts more within the metabolic sphere, at the same time maintaining close bonds with the rhythmical system. Soul life bases itself more on the latter system;

it is more receptive and relates to rhythmical sequence in the pro-creative sphere (ovulation and menses). In the sphere of speech, the deeper incarnation of the male is evident from the fact that the voice goes down a whole octave, whilst in girls it only drops by about one tone.[35]

The rhythmical system has a balancing function between above and below, inside and out. It is therefore best able to confer health. This is probably the reason why women, who live more from the middle, are generally healthier and have greater resistance then men. Yet because women stay more centered and may not step out into the world as much as men do, they may also be less creative in rela-tion to the world. On the other hand females, being centered, go through the stages of development more quickly than males; in childhood and youth in particular they may be years ahead of com-temporary males.[32]

Even before puberty, ego and astral body, having entered into the middle sphere as soul and egoity, advance beyond the metabolism and into the limbs. Penetrating the muscular system they reach the skeletal system by the twelfth year–slightly earlier in girls. This encounter means that the child's thinking now also has 'backbone' to it, i.e. that it can become conceptual and abstract. But the blood also flows in those limbs, acting as the vehicle for the will impulses arising from metabolism. The conquest of the limbs thus has a will component to it. The young person gains a new relationship to the world that is based on will.[36]

On the other hand further penetration of metabolism with ego and astral body goes to the region of the genital organs where the onset of secretory function at the age of 13-15 in males and 12-14 in females signifies sexual maturity.[35] At the same time one notes a 'flowering' of the whole organism that is mediated by the endocrine secretory function of the gonads, and the whole body now becomes distinctly male or female in appearance. In the sphere of the sexual organs this flowering may lead to a fruiting process.

Having helped the organism to maturity for life on earth, the astral body has fulfilled its task where creative physical development is con-cerned and can now partly rise above it. This time new access to the world is gained through the metabolic sphere; it is determined on the one hand by the desires of the astral body and on the other by the will forces that are active in the limbs. The ego is also involved in those will forces, having anticipated developments to some extent from the twelfth year onward in seeking the middle between movements that are sometimes excessive and at other times held in spasm, though it does not itself come to birth as yet.

Emotional life

After its liberation when the child had reached the age of seven, the ether body continued to work in two ways. On the one hand it penetrated the physical body to help it grow and mature, on the other it addressed itself to the soul and established its foundations. In a similar way the liberated astral body now enters into the physical body to give soul to its further growth and impulse to its functions (e.g. the release of spermatozoa and ova, and excretion and incretion in the renal system; the last of these for instance takes the form of resorption of a considerable volume of previously formed urine in the kidneys[37]). The astral body also bears and nurtures the maturing ego; in boys this is more withdrawn into itself, in girls it is more surrendered to the feeling astral body.[38] Thus the astral body is even at this freer level of existence still very much bound up with the physical body, not only stimulating it with experiences gained in the world, but also receiving the powers the physical body has to give. From this point onwards the powers of the various organs influence the astral body in a more differentiated fashion.

This is the situation in which a new form of soul life arises. It originates in the fact that the astral body is not only gaining something–new contact with the world, new powers–but has also lost something–the protection offered by the physical body–for it has now partly come away from it. That protection was provided by the maternal soul cocoon that has been described earlier; it acted as the mediator between the astral body on the one hand and the body and the world on the other. Before puberty, 'the enveloping astral cocoon maintained harmony' (in the soul);[38] after it, soul life developed its polarities. As we have seen, these polarities relate to the lower and upper pole in the human being; this particular development does however have its beginning in the lower part of the organism, and it is from here that the birth of the astral body is initiated and given its special note.

This, then, is the region from which the astral body ascends as it comes free, and to begin with it takes powers from that region with it into that freer existence. The power of desire–a primary power in the human soul that has its roots in the metabolic system and can also turn negative[14]–combines with the soul's experiences and leads to feelings of inclination or disinclination. Drives and passions such as greed, anger, hatred, fear and shame rise up involuntarily, usually with inner or outer excitement. (For the connection between emotional life and metabolism see Ref. 19a (this chapter) and Ref. 18 (Chapter 4).)

It has already been shown how the life of the soul came to rest in the forming of mental images. The principle of the upper human being, the head, where everything comes to rest physically, was coming to expression in this. The withdrawal of vitality from the nervous system went hand in hand with a reduction in movement. In the brain this goes so far that this organ is largely shielded from movement in that it is more or less suspended in fluid. If movement takes hold of it in spite of this, we get the concussion syndrome and unconsciousness.

The tendency to movement in the rhythmical system comes to expression in play. From the seventh year onwards a rhythmical impulse to be at rest begins to emerge in the child's life: Children must learn to sit still in class at certain times of the day, and this relates to the concentration needed for learning. All these phenomena have to do with the fact that the birth which initiates the second seven-year period is from the head.

The onset of the third seven-year period, the period of polarities, characteristically involves the polar opposite to what has been happening until now. In contrast to the principle relating to the head that leads to conscious awareness, the opposite and vital principle governing limbs and metabolism now comes to the fore and grows increasingly more powerful. It is the principle of 'motion'. Rest, the principle essential to health in the head sphere, causes illness in the metabolic sphere. The convoluted intestines are in constant motion; if they lie immobile, like the gyri of the brain, paralysis of the bowels is diagnosed. It is part of the life pattern of ova and spermatocytes that they travel, the ova within the organism, and sperm by actually leaving the organism.

The corresponding soul element is the elementary motion we call emotion, using a term that has the word 'motion' in it. It explains why young people have such an urge in their souls to be on the move, to travel; this is not something wrong but rather the physiological counterpart of birth processes in the soul. Compared to the playful, rhythmical movement impulses of the second seven-year period, the movements one sees at the onset of puberty or even before are more of an urge and have emotional overtones. All such movements have their origin in the desires of the astral body. At the physical level these are the 'unconscious desires' of the astral body in its organic activity: thirst for oxygen, hunger for food lead to the movement of substances and of the juices, i.e. to 'metabolism' (a word based on the Greek for 'change') in the truest sense.[40] The soul may grow aware of the physical desire of the moment, and a close connection with physical processes is certainly evident in sexual desire. Each time such awareness arises the potential for change is also given, and animal powers can be humanized.

The sexual organs are only part of the urogenital system. As the name suggests, this includes both the genital and the renal system. All emotions are connected with the renal system.[26] It has been known for some time that any form of excitement involves secretory processes in the adrenal cortex. Anatomically and functionally the kidneys and adrenals are a unit. During puberty the astral body develops new activity in the urogenital system, where it is dominant; this also takes the form of new adrenocortical activity.[35] After puberty, soul life is initially determined by the urogenital system. If the soul at its core grows excitable and mobile a sanguine temperament develops, and this is connected with the kidneys and the air organism of the human body. A sanguine temperament makes people volatile.[26]

Imaginative Powers and Love

Puberty thus brings the transformation of organic soul powers into more conscious powers of soul that at their lowest level serve to satisfy physical desires. Creative powers are also transformed, but this time not into powers of thought but powers of imagination. Imagination also develops through the transformation of powers of growth,[41] but these are creative powers of growth in the lower part of the organism and particularly the sexual organs. Imagination is an offshoot of the primal power of sympathy, which arises in the lower pole of the organism. It is a full-blooded creative power that rises to enter into soul life and bring vitality to conceptual life.[42] Like the will element that is active in it, imagination looks ahead to the future: 'Ninety percent of twelve- to fifteen-year-olds look to the future.'[43] The wishful thinking that arises from imagination predominates in this.

Younger children do of course have imagination, but in their case it is more determined by the ether body, by the creative powers of nature. Now it has been taken hold of by an astral body that is growing independent. 'True imagination really only comes to birth in the human being with sexual maturity.'[30]

The enhancement and transformation of sympathy is a source of love, and joining forces with the creative imagination this love makes one individual turn to another (not only the sexual partner). 'To end isolation is of course the primal impulse of Eros who lives in creative imagination; after puberty this ... can grow and mature into intimacy between two human beings.'[44] The power of procreation becomes creative imagination in the soul; at its lowest level this evokes sexual fantasies, but it can be transformed into artistic imagination. Just as a new human body is created through physical love based on creative imagination, so something essentially new, a 'creative work', can arise in the sphere of soul and spirit when male and female, two human individuals, unite in creative love borne on the wings of vision and imagination.

This does not mean that imagination and love arise from the immediate transformation of sexual drives, through sublimation of sexual energies, as the Freudian school would have it.[45] The transformation goes much deeper; it harks back to the creative potential of the reproductive organs, and these become creative in relation to something that is outside one's own body. It harks back to the soul potential of a child's love that is still entirely free of sexuality and is released for further development when sexual maturity is reached.

The sexual love that arises at this point is merely one component in the new power of love, for this has the whole world for its object.[36] In women, this power makes the middle aspect that has been so well maintained emerge more strongly and in a new form; in men it results in the middle being more or less won anew. From the middle, which both sexes have in common, develops the entity that encompasses both: the human being. (See the following chapters.)

The power of love can only become free at this stage if it has been allowed to develop in the second seven-year period. Having known authority wielded with love, and having endeavoured to emulate it, the individual now experiences the awakening of love, though during the second seven-year period this is still in its 'pupal' stage.[46] The over-emotional hero worship arising from an excess of sympathy after puberty marks the transition from this stage to real love. Again two extremes may be present: pubescent youngsters may indulge in hero worship on the one hand and be subject to their sexual urges on the other; it takes time until the two can be united in love. We have now again ascended from the lower region of the organism and the soul to the middle region, and this is where we have encountered the power of love before. What kind of soul life develops in this region, to find its final culmination and fulfilment in love?

Sentience and Judgement

Sentience and judgement are two activities the soul strives to develop between the ages of fourteen and twenty-one.

Turning to the world again out of the personal soul life that has come to birth, the individual first of all has the experience of isolation and danger. The protection which the physical body gave to the soul covered not only the inner life but also the soul's relationship to the world. The original existential experience of being 'cast' upon the world (Heidegger, Sartre) now comes painfully to experience in the biography. The paradisaical experience of being naked returns, but this time for the soul: young people feel 'naked' in relation to the world around them; a new hypersensitivity develops and often a hidden shame. The only real way out is to come to love the world.

But it is a long road to that love for the world. To begin with, young people are hypersensitive in relation to the world and withdraw from it into insensitivity, which is made possible by the cool observant faculties of the head. Those are two poles in sentient life; between them lies the space where living, breathing sentience can develop.

Sentience is not a primary element in the soul. It only develops when desires are addressed to the world and meet with perception gained in the encounter, at the soul periphery. When the two have united, 'sensory perceptions have become sentient responses'.[15] Intense perception of a tree leaves me with a certain feeling, and this lends something of its life to the newly perceived memory of the tree. Memory, as an inner perception, also causes the sentient response that has been connected with this to arise again, though this may now take on a different character.

Elsewhere, Steiner put particular emphasis on feeling and will impulses in the sentient response, referring to it as 'feeling imbued with will'. This is in contrast to modern psychology where sentience is considered to arise through physiological processing of external stimuli.[16] The element of will in sentience can be experienced when the latter leads to an act of will. This happens when strong sentient responses arise on the basis of sympathy or antipathy. (I may for instance feel so much sympathy for someone that I get the urge to visit them.)

The connection between sentient response and feeling is immediately apparent. The difference is that sentience is directed towards the world or the inner life, whilst feeling is essentially part of the inner life. Sentience might therefore also be called receptive feeling.

First, however, there is desire, with will impulses arising from it. The soul, which has now become independent, is looking for a new relationship with the world in the third seven-year period. Desire brings about the above-mentioned will-induced encounter with the world. This encounter lives on in the sentient response that has arisen from it. The soul has taken in the sensory perception and made it its own; the fulfilled desire has united in sympathy or antipathy with a content taken from the world. Soul and world have formed a new union in the soul. Yet immediately new desire, demanding new will impulses, arises from sentience of the experience and seeks a further encounter with the world.

Young people live in such experiences and encounters until their late twenties. The life of the soul is like the sea as sentient responses arise in the encounter with the physical world, sink down and disappear, and come up again as memories.[15] One may also think of the sea of blood that fills the organism and brings life to it; sentient responses are the life blood of the soul. The sea of sentience is fed from the springs of desire that issue into the world and take the world into themselves.

Judgement combines with desire from the very beginning. This may take several stages. A desire that rises darkly in the soul for instance, may in the illuminating judgement of the soul prove to be a desire for beauty. The judgement or opinion that 'beauty is to be found in nature', which bases itself on earlier experience, will channel this desire in a particular direction. The desire will unite with an initial perception made in nature, say perception of a distant flowering tree, and out of this a new opinion will be formed, an opinion accompanied by an initially diffuse response: 'This flowering tree is beautiful.' Desire, again guided by opinion, gives rise to a will impulse: 'I must take a closer look!' When that closer look has been taken, the ego has not only developed a differentiated response to the perception but also a definite mental image of the beautiful flowering tree, and this image is based on the sentient response.

Behind the mental image however lies the concept of a flowering tree that initially enabled us to realize that this was a flowering tree and not some other beautiful object. The concept has thus been involved in the process of perception all the way, though it only came to awareness later on. When the concept ('tree') has come alive and grown comprehensive, the idea ('tree') arises.[47] Mental image and concept are already present in germ when the ego half-consciously or consciously forms an opinion.

Judgement always also involves feelings. Zeylmans van Emmichhoven called it 'the other side of feeling that is directed to the outside'.[48] In quite general terms, a feeling of conviction goes hand in hand with the forming of any judgement.[49] When a mental image is formed, this therefore involves not only judgement but also feeling; together with the new sentient responses that have arisen this feeling gives life to the mental images. This life is enhanced if one endeavours to let love enter into one's mental images.[50] The process of forming a judgement or opinion however is imbued with will impulses that are directed towards the goal of this process.

In the process of forgetting, mental images convey something of their life to soul and body; in the process of remembering this life may condense and become essence. The precondition for such life is a living, active faculty of judgement that has feeling and will in it, with mental images and concepts as its offshoots. Individual judgement involves the whole person; it is striven for in the third seven-year period and should be encouraged. It enables individuals to form their own mental images and think for themselves.

At the same time young people can begin to grow out of their subjectivity in forming opinions. Though they have much in common, the connection with the world through judgement is distinct from that through feeling. All living feelings are based on desire for the world; all true judgement strives to know the world through mental images, concepts and ideas. According to Lindenberg, the power of judgement does not ask: 'What do I desire?' Nor does it ask the question that is put when mental images are formed: 'What do I know?' The questions it asks are 'What is true?' 'What is good?' 'What is beautiful?'[51] And the spirit is also involved when these questions are asked in the human soul.[14] The spirit speaks through the ego, of course. The soul's desire that is directed at the world thus has judgement for its companion. In the process of judgement the thinking ego learns to guide the soul. The desire that first rises with dark urgency in a young person is transformed into an alert interest that is to the point, actively perceives the object and continues on into the process in which impressions are digested.

Interest in the world–according to Steiner this is of vital importance in the third seven-year period– is aroused as the 'capacity for judgement' develops; it acts to prevent attachment of the freed astral body to the body and to sexuality.[52] The goal of youthful judgement may indeed be a distant one, yet consciously or unconsciously a course is set for it: In the final instance the goal of one's own judgement is to perceive the essential nature of the object, or of the world, i.e. the heart and core of it that has given rise to it and from which it may arise again in our thoughts.[53]

Personal judgement is like a boat drifting on the river of desire. Young people must learn to navigate these waters. Again and again they will make for a bay where mental images of the world appear as in a mirror, sentient responses arise and the waters in the river of desire then grow less murky and turbulent. Having reached the age of majority, young people have gained the open sea where desires represent the currents, and sentient responses the billows. The light of thought must guide them as they seek new shores, new mental images that condense into concepts and expand into ideas. The more alive the heaving billows of their sentient responses, and the more brilliant the light of thought, the easier will it be to gain new shores. There they can land, gain a firm foothold, take heart and set out for the mountain of true understanding. Yet they may also suffer shipwreck – if gales have thrown them off course, if they have lost their bearings in the dark or if the sea that should carry them grows too shallow. Their ship may sink somewhere along the way if it has not been strongly built.

The Sentient Body

Steiner also called the astral or soul body that is born at around fourteen years of age the 'sentient body'.[1] As already stated, this sentient body, which directly conveys sensory impressions, is closely bound up with the physical body.[54] Speaking of the effect that the arts have on human beings, Steiner even said that the part of the astral body that was united with the ether body stayed in the physical body during sleep. It is evident from what he said that this refers to the sentient body, which then, of course, has a different function from the one it has in the waking state. (It mainly serves to build up the body during sleep). The rest of the astral body–here as elsewhere this means the 'soul'–separates from the physical body during sleep. (See also the chapter on *Art Therapy and the Constituent Principles*.)

Thus it is the astral body in the narrower sense, the sentient body, that is born at around the age of fourteen. Animals also have this, though in a different form. The attribute 'sentient' is important with reference to our subject. Steiner also used 'sentient' as an attribute of the soul principle that develops in the early twenties, thus indicating the significance of this element. 'Sentience', the conscious response to sensory impressions, is of course present earlier on, but it is only with puberty that it develops a more independent life out of the astral body once this has come free. This makes it possible for emotional mobility gradually to achieve a calmness that is nevertheless full of life, and for the static element of sensory perception to be imbued with movement of soul. Sentient life, having come alive in the encounter with the world, is the precondition for 'sensitive inner responsiveness', 'feeling' revolving upon itself, to arise in a centre that always has to be found again.[50]

This has its beginnings in the sentient body, where the ego is preparing for the next stage. Sympathy in spontaneous response to another individual grows into the feeling of sympathy we cherish in our hearts. United with the ego as the centre of the soul, this can become the power of love. A new aspect now arises in so far as this power–which is at a higher level than sentient responsiveness– holds all the powers of the soul within itself. 'All that the ego of man brings to development within him will grow into Love.'[55] In love, desire rising from below and leading to will activity, and the feeling of sympathy from the middle region unite with the judgement and perceptive understanding (no longer just sensory perception) that come from the head region. There can be no love for an object or another individual unless there is sensory perception and perceptive understanding. Desire and sympathy are not love but may become stations on the way to love. The first way in which the goal is achieved is that something of the true essence of the beloved object is more or less consciously recognized and this gives rise to the will impulse to help this essence to present itself more fully in one's understanding or in physical life. When the power of love that is part of one's own nature takes hold of a mental image or an idea, this becomes an ideal to set the youthful heart aglow. This is important in preparing the way for the next soul principle,[56] for out of the middle, the 'heart' of this, the ego will be born, having worked to create that middle during the preceding seven-year period.

The heart is the organ for the power of love. In the discussion of the feeling life during the second seven-year period it was characterized as the organ of balance. The effort to achieve balance, which in the lung lies between inspiration and expiration, becomes concentrated in the heart. As the heart contracts in systole and expands in diastole, the two opposite poles of the organism make themselves felt. In systole, when most of the life-mediating blood leaves the heart, the destructive head pole comes to the fore. In diastole, when the heart fills again with blood and with life, the constructive metabolic pole makes itself felt. Between systole and diastole lie pauses; these create a rhythmical sequence in which rest and movement are united. The ego intervenes in those pauses and establishes the balance between the two extremes. The pendulum swing around a centre creates the basis for a situation where the ego is able to strive for balance and a centre in the soul, so that new elements may come into being.

The sentient body arising from the lower organism seeks its centre in the heart. On the way–in the sphere of the lung–the desirous and emotional life of the astral body opens up to sentient experience by way of sensory perception. This experience can grow into love in the heart. Initially, however, sentient response as well as emotional reaction predominate in the third seven-year period. Sentient response has its basis in the renal system and grows and develops in the soul life of the lung as it encounters the world.

The Fourth Seven-Year Period

Young people reach maturity at about the age of 21. Until then the world influenced them through their home, school and occupational training. From now on they 'face the world as independent, free individuals',[56] and immediately grow aware of the effect it has on them. The period of training and education has come to an end and that of self-training begins, having been in preparation at an earlier stage. Now the 'ego wants to develop the individual's character in free interaction with the world'.[56] The immediate environment no longer trains individuals directly but encourages self-development. Individuals who begin their own development will also be able to take independent action in the world, and this is the point where they may be considered to have come of age. The step from youth to early adulthood in about the 20th year characteristically involves 'the attempt to make one's own decisions, and if the decisions are practicable, responsibility is now taken for the first time.'[57]

These and other phenomena indicate that individuals now have a further essential principle at their disposal. This does not react spontaneously to the world in the way the astral body does when it acts independently. Instead, it applies perception to the world and takes deliberate action out of such perception. Only the ego is capable of this. This specifically human principle is born at the age of twenty-one.

The astral body's schedule of organic development was completed at the age of about fourteen. The ego's schedule of organic development now reaches completion at twenty-one. It had started when the body assumed the upright position, i.e. when the ego made it come upright. The body then grew and took a physical shape that was ultimately determined by the ego. Now, in the early twenties, physical growth is coming to an end: facial growth, and this is where the ego comes most visibly to expression, is complete.[58] It may be envisaged that the ego now comes to itself out of the whole of the individual human form, for its activity in the living body has been comprehensive.

Compared to the view held by many psychologists who consider the ego merely an 'imaginary point behind everything the individual experiences',[59] this is a realistic view of the ego. Buehler, whose researches have been in the field of biography, states, 'that in the stream of changing events and of all the transformations the individual must go through, there is a fundamental core content.' She calls this the 'self' and considers that all progress in the biography comes from this.[60] Jung's 'self' includes not only the conscious ego but also the unconscious sphere with nonpersonal contents such as the archetypes of man and world. The ego concept expands, and soul development is towards 'self' in the process of individuation.[61]

As already mentioned, Steiner, too, put the ego at the centre of the soul. But it also acts on the living body via the unconscious soul life, and on the other hand opens up to the light of the spirit. It is not merely part of essential human nature but also part of the world, and ultimately the spiritual world, with the higher ego remaining in that world and continously irradiating human existence. Initially we perceive only the 'reflected radiation' that is the lower ego, which is our earthly ego.[62]

All human development depends on the individual personality opening up more and more to the radiance of the higher ego, so that more and more of the higher ego enters into the emanations of the earthly ego. The heart is the physical organ in which higher and lower ego are united and from which the higher ego then radiates into life.[63] Love felt in the heart brings this to realization when it reaches its highest development, and ultimately this love derives from the higher ego. The true essence of the person with whom the higher ego wishes to unite in love is the higher ego of that person.

Once again a polarity emerges. The ego experience is an inner one: as it develops, the individual strives to make it more and more the centre of the soul. On the other hand, the ego is the lodestar that guides soul development. This was called a person's 'genius' in antiquity, a guiding spirit hovering above the individual.[64] In taking up a challenge, individuals 'grow beyond themselves' and towards their genius or higher ego. During sleep, the lower ego separates from the living body and unites with the higher ego that dwells in the spiritual world. In the morning, the lower ego may be sentient of some of this encounter and bring it into daily life in the form of new insights and new strength.[65]

During the day, individuals will first of all feel separated from their higher egos. They realize that during the day, too, the life of the ego has both centre and periphery and that this life lies between self assertion and the giving-up of self.[66] The archetypal image is as follows: From the periphery, and ultimately the spiritual world, the ego shines into the inner life, which is closed in upon itself, and from the centre of this radiates back into the world and ultimately the spiritual world. The ego is able again and again to create a balance between the two extremes, to establish a middle region. When we give ourselves up to the world the centre of our own essential nature can be made part of this; self assertion can also include the periphery, the other individual. This alone allows love to arise in the ego, a love in which the individual personality does not give itself up in opening up to the other but unites with the other in freedom and finds fulfilment.[67]

The most central expression of the way the ego 'breathes' between point and sphere lies in the living experience of the heart. Systole and diastole mediate not only between above and below but also between inside and outside. The physical basis for this is the fact that the inner heart is connected with the outside world through the blood that streams to the lung and comes from the lung. The one enters into the other, and again we have visible evidence of this, for contraction of the heart goes hand in hand with expansion of the blood as it leaves the heart, while expansion of the heart means contraction of the blood as it streams into the heart.

The more strongly and all-inclusively the soul is 'breathing' in the heart, using the physical processes as a basis, the stronger and richer grows ego-based love, with centre and periphery responding to each other in the way that has been shown above.

Love enables the ego to unite with other egos that come within its horizon. The following diagram may serve to indicate this.

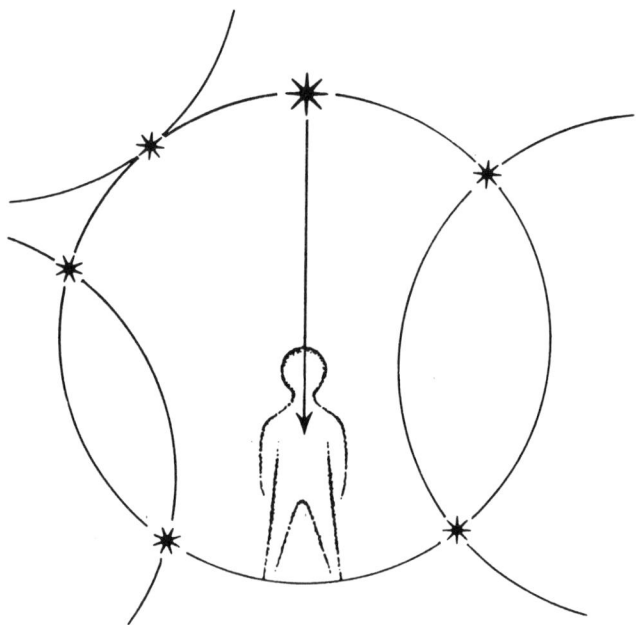

When the light of the higher ego has entered (vertical arrow) the earthly ego of the individual is radiant during life on earth (not shown in the diagram) and this radiance forms the ego circle, the ego's horizon (see diagram). Other ego circles may come within this sphere of radiance for shorter or longer periods. Sometimes they will merely come up briefly on the horizon (see tangential circle and point top left). Potentially those tangential points or points of intersection can become stars if the higher ego of a partner penetrates to them. Such encounters are quite rightly considered starlit hours in human life. In the diagram, only the first and last encounters in a longer-lasting relationship are marked with a star; those are points where the higher ego of the partner and one's own may shine forth with particular brilliance.

Thus biographies combine in harmony, and this usually begins to happen after the ego has been born in about the twenty-first year.

Preparing for the Birth of the Ego

Before the ego can be born it has to find itself. It is understandable that at this stage the emphasis is on centre-seeking ego activity and that this can only gradually open up to encompass love for others.

A predominantly egocentric attitude is normal in children and during the earlier part of the second seven-year period. Young people entering into puberty need the maturing ego as something to hold on to in the chaotic experience of puberty. That experience is something every human being has to go through, and some will be more affected by this than others. The concept of identity–a term one often hears today–comes up, and the aim is to identify with one's self, i.e. a state of 'constantly being one's own self'. Earlier, the young people identified with others, now they are feeling their way to 'ego identity'.[68] The ego has set out on a road that leads to the centre of the soul, to the centre of calm in an ever-shifting soul life.

The first signs that the astral body was coming to birth appeared in about the twelfth year. In the same way the first signs of the approaching birth of the ego show themselves in about the eighteenth year. The early upheavals in soul life have settled, the choice of occupation that has been made of one's own accord or under pressure has its effects. Adolescence has started. According to Lievegoed this is a transitional stage beginning at the age of sixteen or seventeen and continuing until the young person is between twenty-one and twenty-four. The 'central issue' at this stage is 'Who am I?', 'What do I want?' and 'What am I capable of?'–questions concerning the self that indicate that the ego is about to come to birth.[69]

Glas and Lauenstein in particular have shown that at this point the rhythm of the moon nodes, i.e. the progressive points of intersection between the paths of the sun and the moon, prevails over the seven-year rhythm in the biography.[70] 'Every 18.6 years the moon nodes are in the same position in the heavens as at the birth of the individual.' A new birth situation arises every eighteen years, seven months and nine days, and at these points in life individuals are not merely emerging from their previous state but also aware of something new approaching in life.[71]

The birth process that occurs every seven years is therefore given a special note every eighteen or nineteen years. At the age of eighteen, a stage begins where physical development comes to a close and the birth of the ego approaches. At the age of thirty-seven, which is the next moon node in life, the middle of life is passed and preparation made for the birth event that signals development in mind and spirit for the early forties. At fifty-six, with the third moon node, the last of the nine seven-year periods begins, and this is particularly heavy with destiny. In the seventy-fifth year, the time of the fourth and usually last moon node in human life, the individual has passed the biblical age of seventy-two and entered into old age. A new door opens on to further progress in the earthly or into the spiritual world.

Steiner spoke of the 'Door of the Moon' by which a human being enters into earth life and of how individuals are determined by the past at that stage. They can now set out on the road to freedom and take the road to the 'Door of the Sun' that will finally let them pass into the world of the spirit.[72] It is probably fair to say that in about their eighteenth year, at the time of the first moon node, human beings pass through a lesser Door of the Moon. The door to the next seven-year period, which will be in the sign of the Sun, is opening.

Birth of the Ego and Soul Development
Soul development started with the birth of the astral body in around the fourteenth year. Initially it was still in the shadow of physical development, however. In the process of coming free the astral body had to enter into that development and come to terms with it. This is also evident from the close connection between the astral, or sentient, body and the living physical body at that stage, with even the term 'body' indicating an element of form or shape relating to physical

development. The third seven-year period also belongs to this stage of development, for it is only after this that the first seven-year period of soul development can begin.

The first dawn of soul life showed around the fourteenth year. In about the twenty-first year the ego sun rises in soul life. From now on it will shine on the whole of soul development. It is part of essential soul nature that the soul develops through the ego and not out of itself. Like the ether body, the soul is a middle principle; it unfolds between the living body and the spirit. From about the age of fourteen, when it has become more independent of the body, it can gradually perceive and recognize the sphere of the spirit to which it hopes to rise as it develops. It is, however, through the ego that the spirit lives in the soul, and the impulse for soul development therefore comes from the ego. In animals, who have only an astral body and no ego, puberty means the end of development; in humans it means that preparation begins for the birth of the ego. Once animals have developed the capacity for reproduction and undergone the physical changes that go with it, nothing really new happens in their lives. The animal 'soul' as astral body remains closely bound up with the physical body. In humans it begins to separate from the physical body at puberty. Once maturity for life on earth has been reached, soul development begins on earth–initially still in the domain of the living body.

The birth of the ego in about the twenty-first year again marks not merely the emergence from an old state, i.e. from the diffuse kind of life that the ego had in the soul. In the three years preceding its birth the ego caused something new to come into the young person's life, and this came to be connected with the experience of choosing one's occupation and taking the first steps in that direction. The encounter with the opposite sex also brings awareness of a third element, one that wants to unite the two partners physically or at the soul level. Responsibility for the unborn child and for common goals may develop and in some cases lead to an early marriage that may be consciously desired or not. Eros becomes love,[73] and not only are there desire and the enjoyment of shared pleasures and of feelings of sympathy, but the partner's true nature is divined. It is now possible to set out on the road from 'I' to 'you'.

In a wider context individuals are now more consciously looking for a group, though this did also play a role in the third seven-year period. The group stands for a part of the human environment with which one feels in sympathy and also has common goals. The world is experienced through the medium of the group, preferably the peer group. In the present age, 'fitting in with one's own generation in society' is the goal of young people in their twenties.[77]

Glas gave a number of examples that characterize the preliminary stage in the birth of the ego at around eighteen and also the actual birth.[70] Among other things these show how important it is for the will that is coming awake and growing more clear-cut as individuals learn to make their own judgements, to be given the opportunity to become active in the world and in response to the world. The will of the ego initially receives its impulses from desire and comes into effect compulsively; it often needs opposition to make it come awake and grow strong.

The life of the German philosopher poet Friedrich von Schiller (1759-1805) provides a good example. He was eighteen and a student at a military academy when he started work on the theme that was to become his first play *The Robbers*. His character also changed at that time and he grew 'more self-confident, more bold'. He did not write the play until he was twenty-one; then his newborn ego gave expression to one of his life's ideals: the struggle to achieve freedom from tyranny. His own true aims had been suppressed at the academy, and out of this arose the impulse to create something that was of general validity, his ego's horizons having expanded with its birth. The work first matured within the peer group, however, for the dramatic play about the captain of a band of robbers who had their lair in the woods was first read to Schiller's friends in the woods near Stuttgart. A second ideal arose in Schiller's life and poetic work and was experienced and given form by the ego: the ideal of friendship.[75]

A basic law of biography is also demonstrable in the subsequent life of the poet. Opposition and the overcoming of opposition are as much part of human biography as the gifts and the acts of renunciation that were characteristic of Goethe's life. What would Schiller and his work have been had it not been for the opposition from the Duke of Wuerttemberg and the obstacles presented at the academy? If Schiller had remained a regimental surgeon, would he have achieved his life's mission and brought the impulse of his ego to realization? What would Goethe have been without the gifts life had given him, gifts he had to take up positively or renounce? In Goethe's case it was important that he resolutely entered the carriage the Duke of Weimar had sent and went to live in Weimar. Schiller on the other hand had to set out on the journey that was important in his life behind the duke's back; he thus overcame the obstacle in the path of his youthful destiny and gained the freedom he needed to reach fulfilment.

The ego sets out to fulfil its destiny in a living body that has its hereditary traits, which derive from the parents. Out of living experience of the imperfections and faults of a previous earth life it has chosen its own destiny–Steiner has spoken of this many times–in order to go through a process of further development on earth. The destiny that apparently comes from the world around the individual has also been chosen; it has been drafted like the main themes of a play that is then fully worked out in coming to grips with the material offered by life on earth. Physical birth places the ego in an environment that is like an enlarged physical body, its 'social body'. When the ego is born out of the physical body at about the age of twenty-one and becomes an independent principle, it incarnates in its 'social body' and relates to the environment through action and reaction.

Human beings differ from animals in that they do not 'adapt' *to* the world but develop *in* the world. They are not at the mercy of the environment the way animals are. An astral body remaining in the physical body shapes animal life in conjunction with the animals' group ego. The long period of youth during which the human individual is relatively protected serves a vital purpose. During this time the young are able to come to themselves before they come to grips with the world as individuals. The early years create an area of freedom that can then be filled with life by the adult.[76]

The Sentient Soul
The ego's creative potential is evident, among other things, from the fact that the ego is born with one of the fruits of its earlier activity. The twenty-first year marks a special event in the biography because two births occur at this point. Apart from the ego, and at the same time also through the ego, the sentient soul is born as the first principle of soul development.

With regard to the use of the term 'birth' in the context of soul development, it should be noted that Steiner, on whose ideas this is based, only used the term 'birth' with reference to the physical, ether and astral bodies and the ego. According to Steiner, one of these undergoes 'particular' development during each seven-year period after the twenty-first year, just as they did during the early seven-year periods.[56] Elsewhere he also used the term 'birth' with reference to the soul principles,[77] an image that seems true if one consideres the birth pangs of the seven-year periods that lie ahead. The term will therefore continue to be used.

With regard to the preparatory work of the ego that leads to the sentient soul being born in about the twenty-first year, Steiner said quite generally that this is not a rigidly predetermined sequence where work on a particular principle is only done during the preceding seven-year period. Preparation proceeds 'from the first flicker of the ego'. This is also when the 'transformation' of previously established forms begins, making the old form the basis for the new.[1] The ego has therefore been working to create the preconditions for soul development during the three seven-year periods that were essentially devoted to the physical body. One can see why this is so, for any development in soul and spirit has its basis in certain preconditions being given in the physical body, where the ego has been active from the beginning. Nevertheless, specific preparation for the sentient soul is made during the third seven-year period. Part of the sentient body is transformed and as a result the first soul principle arises; this happened in human evolution and now happens in the life of the individual.[78] The struggle the ego had with the polarities and the inner chaos of puberty is now bearing fruit; the sentient soul is born, and undergoes its development between the ages of twenty-one and twenty-eight.

Once again sentience gives a particular note to a seven-year period. As described earlier, it arises as desire and judgement unite with sensory perception. This time, however, the ego plays a greater role, and the soul life of a young person, which continues to be swayed by emotion, becomes more centred. One gets the impression that the ego either gives itself up voluntarily to emotional turmoil or steers its way through it. The 'boat' of judgement that went out to sea and proved its seaworthiness in puberty now has the ego entirely in control at the helm. During the third seven-year period the helm was occasionally given to other individuals to whom the soul felt drawn; these helped the soul to develop its own judgement; it would not follow them the way it did during the seven-year period of authority (seven to fourteen) but rather let them help develop and confirm its own judgement. Now, in the fourth seven-year period, the ego follows the same course as its peers who have formed similar judgements to its own.

A fundamental psychological difference is seen between the sentience of the sentient body and that of the sentient soul when we consider the memory process. This process had already been given greater emphasis during the time when the sentient soul was in preparation. The sentient body, which humans have in common with animals, reacts instantly to sensory stimuli. The sentient soul produces memories in a process based on ego activity, and these form the background to all experience and all actions taken.[79] In the case of the sentient body, therefore, the direction is from the outside to the inside, while that of the sentient soul is from the inside to the outside.[80] Animals are passive in their vision, opening their sentient body to anything that streams in from outside. The sentient soul lets the eye look actively to the outside. The sentient body causes the eyes of the young person to be opened to a new vision of the world at the beginning of the third seven-year period, and this may in the extreme case become a passive stare. As the third seven-year period progresses, the emergent sentient soul directs the vision of the struggling ego to the world outside.

The sentient responses arising from the ego aspect of the sentient soul again put increased emphasis on the emotional life of the middle region. In contrast to the veiled emotional life of the second seven-year period, however, the ego relates directly to the world through feelings and sentient responses. 'At this point feeling becomes destiny.'[81] The ego enables the feeling of love to become creative, sentient responses are more aware, and judgement leads more immediately to mental images that bring the enlightenment which is the goal of the ego. Mental images and ideas that have already become ideals are now taken into the world by the ego. Experiences gained in the world may cause them to be transformed so that they fit in with its laws. The basis is again provided by the memory process.

On the other hand it should be stressed that the ego response, too, is immediate, and sentient responsiveness continues to be the dominant element in the fourth seven-year period. This element has to be constantly re-created; the sea of sentient responses needs to be fed anew by the rivers of desire and the images presented by the world. This inevitably means turmoil and rough waters. Everything 'we experience by way of joys and sorrows, pleasure and pain, drives, desires and passions' lives in the sentient soul, in short everything 'that comes to life in the soul through direct stimuli supplied by the world of sensory perception. The ego has not yet woken and is not yet wholly there.'[78] Emotional storms are calmed in the ocean swell of sentient responses, but they continue to bring movement into the waters, keeping them alive.

The sentient soul also uses the urogenital system as a substrate. Now, however, the 'sentient experience' of this organic system belonging to the astral body[82] ascends much more strongly from the emotional life of the lower organism to the mid-region of new feeling, to the lung and the heart. The urogenital system becomes an instrument of soul development once the astral body has partly left it during puberty so that the ego is able to relate to it more freely. Yet sympathy and antipathy are still the polar principles that determine soul life; a lesser form of these are the inclination and disinclination that have their organic substrate in renal incretion and excretion and have their counterpart in sentient responses or feelings in the soul life of the lung.[26]

The connection with the lower region of the organism also comes to expression in body language. People who live wholly within the sentient soul like to pat their stomachs.[56] It is popular to keep one's hands in one's pockets and this points in the same direction, especially if the pockets are over the abdomen.

The ego aspect of the sentient soul is clearly at the helm, but it does not yet have firm hold of it. It therefore does not imply pathological disorder if the helm temporarily slips one's grasp. Considering soul life as a whole one does discern the ego at the centre, but the spark is only a dim one that glows intermittently. The light of the ego is given up to the ocean swell of the soul, and the ego to the periphery of soul life where it is able to meet the world in inward experience.

The vital question for the sentient soul which becomes the question of destiny for the fourth seven-year period is: 'How do I find the world and through the world my self?'

The Fifth Seven-Year Period

In the late twenties, at about twenty-eight years of age, the character of soul life tends to change again. The turmoil and upheavals have gone and individuals seem more self-contained at this age; at the same time their actions are more thought out and decisive. They tend to reflect more on life and its conditions where before they took them more or less as they came. They are inclined to keep their sentient perceptions alive in the soul and let them mature into feelings.

This new situation also means a new attitude to life in the world. 'Almost all of them want to change their lives,' is Sheehy's summing up of the new phase.[83] It is not uncommon for them to change the occupation they originally chose for themselves or had chosen for them, or they see their work in a new light. Quite often is is only now that 'the final and life-long occupation is chosen'. Relationships also change, and this applies to both inner and outer relationships. In the USA previously contracted marriages most frequently came to an end when the husband was thirty and the wife twenty-eight.[83] This is much the same in Europe. Communes complain that young members lose interest in communal life and withdraw when they have reached the age of thirty.

Lievegoed calls the fifth seven-year period the 'organisational phase'. 'One's youth is over and life is now getting serious.'[73] People are setting their sights on practical goals, with common sense determining the route to their achievement. Considerable organisational skills are also demanded of parents. The children of couples who married young are now going to school, the household needs more careful organizing than before, and parents want to, or have to, keep up with their children's lessons. Marriages contracted in the late twenties are of a different calibre and are more governed by common sense than previously. This also applies to the care of any children born to these parents.

In marriage and in human relationships generally the aim is now the kind of inner development that has already been mentioned. Now people are not simply drawn to people but they also want to take the other person into their minds and hearts. When they come to love someone they do not merely have an intuition of the essential nature of the other person but try to nurture and cherish this essential nature in their hearts, just as they seek to keep their encounters with the world alive in thought and feeling.

It is evident from the above that the ego is no longer mainly address-ing itself to the world and living at the soul periphery. An inward road is taken in the late twenties. The state of 'inner equilibrium'[73] that comes with ego-based inner development is then also achieved in relation to the world. During the fifth seven-year period, inwardly the calmest phase in life, individuals can enter into rhythmical, 'breathing', exchange with the world and find themselves in equilibrium in the process. At times they will intervene in the world and organize it, at other times they withdraw from the world to reflect on it in mind and heart. Mind and heart however also radiate warmth, whilst the rational mind throws light on the inner life, where 'con-scious awareness' can begin to 'gain understanding of itself' and methodical discovery of self (and self development) may be initiated.[18]

This turning point is also reflected in Schiller's poetic work. It came at the time when he wrote his play *Don Carlos*, a drama of ideas show-ing the transition from the passionate emotions that fill the sentient soul of the Spanish Infante Carlos to the new state of soul that is exemplified in the figure of the Marquis Posa. 'Sire, I ask you to grant freedom of thought!' These words are characteristic of a new thinking that lives in, and is given utterance with, warmth of soul and now shows the way to human freedom in a much more explicit way. This was the first play Schiller wrote in blank verse. It shows the increasing desire for form that derives from the thinking pole in man and was later specifically described by Schiller.

In Goethe's biography, the transition from the fourth to the fifth seven-year period is distinctly marked. At the age of twenty-six he followed the call of the young Duke Carl August and went to Weimar where he initially continued in the style of his *Sturm and Drang* period together with the Duke. A gradual change followed. Just before his twenty-seventh birthday he became a government official and developed an interest in the mining industry, in geology, mineralogy and later also botany. At the same time an inwardness of heart and soul developed and this came to destiny-determined fulfilment in Goethe's love for Charlotte von Stein. The influence of this woman brought order and clarity and made an essential contribution to Goethe's development over the ten years that followed.[84]

The key to this stage of development is 'order' in the widest sense, and the idea of order is also contained in the Greek word *cosmos*.

In Goethe's life the impulse of this new epoch ranged from finding his place in a cosmos that was continuously expanding for him, encompassing the world of nature and that of the spirit, to establishing order in his daily life. Two passages taken from letters written by Goethe demonstrate the difference between the fourth and fifth seven-year periods in his life.[85]

The first was written when Goethe was twenty-three years of age. 'Out on the tiles last night, this morning driven out of bed by projects. The inside of my head looks just like my room–not a piece of paper to be found except for this scrap of blue. But any piece of paper will do to tell you that I love you.'

The second letter was written nine years later. Among other things it says: 'Keep my letters in their proper order from now on. Perhaps you could see to it that they are put in a file, just as I shall do with yours, for time passes and we had best let the few things that remain to us increase in themselves by letting order and firm decision prevail.'

The turmoil of *Sturm and Drang* may come to different expression today and the sentiments in which it comes to expression may not be the same. Fundamentally the situation is still the same, however, and this applies also to the inner attitude for the fifth seven-year period that came to expression in the second passage.

Once again the new soul life is utterly different from the one that went before, and this in itself is sufficient to make us speak of the 'birth' of a new soul element. Steiner called the new element the intellectual or mind soul.

The Intellectual or Mind Soul
The dual term suggests that this soul element does not consist of two parts but develops mind and intellect as a whole. (The same holds true for 'intellectual and mind soul' and 'intellectual mind soul'.) The emphasis may be more on one aspect or the other, but mind and heart are always involved in thinking, and thinking also has its role in mind and heart. The intellectual aspect needs warmth of heart if it is to relate to life. Thinking plays a role in the mind and heart aspect from its very inception, with reflection bringing calm lucidity into the fluctuating life of the sentient responses, passions and desires of the sentient soul.

The thinking that leads to judgement in the sentient soul is still guided by sympathy and antipathy in its search for the truth. The new soul element seeks to go beyond subjective attitudes and reach the objective truth. The spiritual reality of this truth is not yet grasped, for sympathy and antipathy also influence the intellectual soul, but truth is already proving attractive to the soul. It determines the further development of the intellectual or mind soul which only now lets the independent thinking of adulthood become fully apparent.[86, 6]

Reflection on the element of truth in an encounter or experience depends on the individual letting the 'external stimulus continue ... in his inner life.' If we do not merely let perceptions 'come alive again in the sentient soul but reflect on them, if we give ourselves up to them, if we gain new experiences, they will grow and take form as thoughts, judgements, and the whole content of mind and heart.'[78]

As already mentioned, the ego has grown stronger and more active at this stage, and its increased activity measures up to 'life in all its seriousness'. The new soul element has in fact arisen out of this activity.

The genesis of the intellectual or mind soul, once again the consequence of transformation wrought by the ego, occurs at a deeper level than the genesis of the sentient soul. The thinking and the mind and heart aspects of the new element relate not only to the soul. They appear to have greater permanence than the rapidly forming and changing sentient responses. We are here dealing not only with soul powers but also with capacities that have their roots in the ether body. These are the capacity for independent thought and the capacity to take things to one's inner heart, both of them potentially available in the late twenties. Light is thus thrown on Steiner's statement that the intellectual or mind soul arises through work that the ego does on the ether body.[78]

This establishes a connection to the second seven-year period, the time when the ether body developed. During that period, when children were not yet able to discover the truth for themselves, it was important that there was authority wielded with love that would be seen to be the vehicle of truth.[87] That is how the foundations were laid in the ether body so that later the transformation of creative powers into thinking powers could be enhanced and an organ for the perception of truth developed. Yet the growing inwardness that comes with the fifth seven-year period also has a connection with the second seven-year period, for the experience of isolation around the middle of that period, at about 9 or 10 years of age, is potentially the hidden germ of an inwardness that is to come at the later stage.

Relationship to the Physical Body

The relationship to the world changes, and so does the relationship to one's own body. As it withdraws into itself from the world, the ego also loses its immediate relationship to the living body through which it relates to the world. The end of one's youth is also felt in the body, for now it no longer has as much to give to soul life as it had before. We shall see how the rising curve of vitality begins to level out and the body consolidates. This means the loss of the elemental powers that the ego was able to gather from the vitality of the living body. The consolidation of the body is no longer the source but the foundation of the intellectual or mind soul. This soul now goes its own way, which is no longer the way of the physical body.[88] Consolidation does not mean growing inward, but the process may use the consolidation as its base.

At the organic level the intellectual or mind soul no longer has the urogenital system for its base–mention has been made of the intense relationship between that system and the world–but the liver and gall-bladder system, the functions of which relate to the inner organism.[26] The liver, the central organ of the water organism with its own internal circulation, is not dominated by the ether but rather by the astral body. The ether body creates form out of the watery element; from this point on it serves the life of the soul more than it did before. (In earlier years, the watery element did provide the base for a developing phlegmatic temperament.) The biliary system, which is present everywhere in the liver, is the means by which the ego radiates into all the vital functions of the liver and thus gains the power of will. (The choleric temperament has its roots in the biliary process; it is a will-dominated temperament, and this has to do with warmth being enhanced to become fire.[26]

Compared to the sanguine or emotional 'kidney person', a 'liver person' has a phlegmatic temperament with warmth of feeling and quite a marked will element. This, however, is merely a more extreme case of something that happens in every individual at this stage, when the hepatocystic system becomes the organic base on which the intellectual or mind soul develops due to the work that the ego does on the ether body. The biliary system enables the ego to radiate not only into the liver but also into the ether body, the liver being the chief organ of the ether body. The biliary system provides the base for greater will activity on the part of the ego; as a result the inner circulation of the liver's water organism becomes the inward pondering in mind and heart.

The evolution of the intellectual or mind soul also involves another organic region. The sentient soul has the urogenital system for its base but grows and develops in the life of sentient response to the world in the middle region of the human being and especially the lung. The intellectual or mind soul bases itself on the liver and gall-bladder system and in the middle region enters into the life of the heart which is now coming much more to the fore.

You 'beat your breast' when the life of the sentient soul rises up to influence the intellectual or mind soul. The same idea underlies the expression 'with all my heart'.[6] The heart has the central role in the process of growing inwardness. So far the head has been taking things in and sentient responses have arisen out of this. Now 'knowledge of the heart' is to arise.[89]

To ponder things in one's heart is to keep them alive in the highest sense. 'But Mary kept all these things and pondered them in her heart,' it says in the Gospel of St Luke (2.19). That is pure mind soul, a heart open to the divine spirit.

The vital question for the intellectual or mind soul may be formulated as follows: 'How can I perceive order in the world, and see my own life as part of that order?'

The Sixth Seven-Year Period

In a life span of seventy years the middle of life is marked by the thirty-fifth year, in a life span of seventy-two years by the thirty-sixth year. Taking a wider view it is also possible to say that the middle of life is between the thirty-fifth year and the early or middle forties. This makes the sixth seven-year period the middle period in life.

Extensive researches have shown that humans reach a peak between thirty and forty years of age, with the maximum somewhere in the mid-thirties.[17] All being well, people have grown capable in their thirties. The undertakings of the twenties have been brought to realization, and individuals have now achieved status in the world and are aware of this. Increased self confidence goes hand in hand with conscious experience of having got there by using one's own will, and initially the feeling tends to be that things will go on in the same way.[73]

Similar feelings of elation are often experienced when the top of a mountain is reached. Something has been achieved. But then we look around–usually first of all back along the route we have climbed. So the peak has been reached; but how was the chosen route? Perhaps a better one could have been chosen. Why did we set out in the first place? What had been the expectations, what kind of prospect did we hope for? And how do we feel now that the peak has been reached–has it been worth the effort?

The question as to the value of what has been achieved arises when we look back from the mid-life peak. During the sixth seven-year period it becomes something we demand of ourselves: 'We do not just want to have done the right things; something of value should also have been gained.'[90] The question as to the meaning of life, which had already come up in younger years, thus becomes the question as to the concrete value of life. A young person would ask: What is the meaning of life? An adult would ask: Has my life been of any value so far?

After this we look ahead from our mountain top to where new goals lie in a distant haze. Will it be worth it, and do we still have the energy to achieve them? In mid-life human beings face their own death. 'Young people never consider the fact that they, too, will die one day.' Certainty of our own death generally only comes in mid-life.[91] The crises and disorders that arise from the experience will be considered later. To begin with, we shall consider its consequences.

Another question that arises is: How far does the horizon that I see before me extend? How many years of life may be left to me? This question may develop into a challenge to look for the best way that will take us from the mid-life peak to the goal that lies before the horizon. The way is determined by the question as to what value one's life could still be to the world. And the question as to the value may then become the question as to one's mission. Now that the peak has been reached this mission will be seen more clearly or indeed only for the first time. And suddenly we have the feeling that something is coming towards us, more so than we were able to perceive before, or it may indeed be something quite new.

The individual finds that the peak is not merely a high point but that it has opened up a new world. The eye looks back and ahead and also upwards. The experience is that the new element that wants to come into one's life, the mission that so far was an unconscious urge, a conventional or perhaps ideal goal, or even not envisaged at all, does not arise out of oneself, nor does it have its origin in what lies ahead. It comes 'from above' and unites with everything that has been growing towards it as life progressed and which is now seen to be the true mountain peak in the life of the individual. The final mountain peak experience becomes the experience of a new birth, and the soul has opened up to this on the mountain peak. Growing beyond themselves, individuals can come to experience a higher principle and give themselves up to it, uniting with it by taking up their mission in life. In the final instance this higher principle is a ray from one's own higher self or ego; through it, a new element arising from the world of the spirit wants to come to realization in the life of the individual. (This aspect has already been referred to in the discussion of the idea of 'genius'.)

The peak reached at mid-life also marks the end point on a path taken by the ego. We have seen that this path was first taken in about the 28th year. It led to the inner soul; individuals have gradually become able to act more and more effectively out of their own centre and influence the world. If accord is to be reached between inner and outer life the ego clearly must reach the centre of the soul by mid-life. Again a target has been achieved. The last time the ego was able to come free of the living body; now it can come free of the soul, having taken its path through it. The ego's experience of the soul peak leads to a more direct confrontation between ego and spirit, and the higher ego then shines forth from the sphere of the spirit.

This enhanced awareness of the spirit, and of one's own spiritual nature, combines with enhanced awareness of living in the world and of the world itself. As the ego emerges from its soul cocoon in mid-life, human beings grow more awake and develop greater awareness in all directions; once this point has been reached they enter into maturity. 'Inner maturity can only be achieved in individuals who live in awareness.'[92] Considering this situation we can also see why Rudolf Steiner called the new soul principle the 'awareness soul'.

The Awareness Soul

Born at around the midpoint of life, the intellectual soul develops during the sixth seven-year period. Humans first achieve bright daylight awareness in mid-life through the senses, with the ego taking in the world of the senses in a new way. People begin to 'come out of their shells again.'[78] The intellectual or mind soul brought a breathing process of growing inwardness. Now individuals relate more intensely to the world again, as they did earlier when the sentient soul was dominant. But the sentient soul united human beings with the world; now they observe the world more carefully. If they do not limit themselves to the superficial aspects of the sense-perceptible world, the quest for values that we have already spoken of will begin.

In the early thirties, individuals were reflectively moving within a specific value system that had been given by the world. Their values may have been material, or they may have related to higher principles such as truthfulness, justice, beauty. Depending on character and one's position in the world, preference will have been given to one or the other. 'Values are preferences'[93] according to Buehler, and up to the middle of life this may well be true. Then, however, people ask themselves: Why do I give preference to one particular value or another? What lies behind this? Have I been motivated by a desire to be taken seriously? Beauty has made life precious to me so far, but what is its real value? What, indeed, is 'beauty'?

Values are now seen against a background that goes beyond preference. This makes the question of values into the question about the essential nature of things. Questions as to the essential nature of the world also give rise to the question of one's own essential nature. Discernment of the object, of its innermost essence, the spiritual element in it, formerly sought in sentient response, in thought and reflection in the soul and pondered in the heart, now shines forth in the soul's centre, in the ego. The ego alone is the spiritual principle in us that is able to answer the question as to the essential nature and spiritual core of that object.

The awareness soul lives out of its centre when 'the eternal' shines forth within it.[86] This happens whenever the essential nature of something is recognized. Then the higher ego irradiating the awareness soul unites more closely with the lower ego and reveals to it the essential nature of that object. Such recognition is the fruit of a new way of thinking. The reflective thinking of the intellectual or mind soul gives way to projective thinking through which the object is recreated from within, out of its own essential nature.[6] With this projective thinking–Goethe called it thinking based on reason rather than intellect–human beings unite with 'the world's creative wisdom'; the will, which reaches out to the future, influences their thinking. It is the will that provides the most powerful impulses in the awareness soul.[6]

In the sentient soul, feelings for the world were the dominant element. In the intellectual or mind soul, the emphasis was more on thinking, and feelings became more inward as mind and heart developed. In the awareness soul, the will activity that so far has been bound up with thinking and the inwardness of mind and heart grows independent and addresses itself to the thinking of the intellectual soul. Thinking things over thus becomes thinking ahead (projective thought), and at the same time will impulses begin to serve the realization of goals that have been recognized in the spirit. Thinking comes to life in a new way in the awareness soul and it also comes much more awake; will activity is purified in the light of awareness and gains in power in the process. Human feeling emerges from being held close in mind and heart and grows aware of love for the true essence; this love combines with the experience of a new, spiritual freedom.

At the centre of the awareness soul, individuals recognize the essential nature of the world; their own awareness comes alive in the process and thus finds the way to self awareness. The new relationship to the world does not depend on the ego entering into the soul's periphery the way it did before, but on its following the higher ego out of the centre of the soul to win through to a new, spiritual relationship to the world. This is a process of turning inside out, and we shall come back to it in the final chapter. The principle that recognition of the world and self realization have in common now shows itself to be the world of the spirit; this is alive in the earthly world and in human beings alike. Human beings are now able to see themselves as members of that spiritual world who are endowed with conscious awareness and come to see the world in a new light.

The Awareness Soul in Mid-Life

The fact that in mid-life human beings find that there is something greater than themselves within them, was put in the following words by the actor Joseph Kainz at the age of thirty-five: 'I have found something in me that is not a part of me; I am part of it; I am going to nurture and develop it; I want to be like it. I want to enter into it with all my being.'[94]

Goethe experienced both sides of spiritual realization in his sixth seven-year period. 'It was in Rome that I first found myself,' he said in retrospect.[95] The experience of looking at classical works of art moved him to say: 'There you see necessity; there you see God!'[96]

The two statements come together in a third: 'I want to procure eternity ... for my spirit.'[95]

In the years preceding his Italian journey–he was thirty-seven when he started on this in September 1786–Goethe had been deeply involved in botanical studies, very much in the style of the fifth seven-year period. Following Linnaeus' line of thought, he considered the principles by which the world of plants was ordered, and familiarized himself with botanical systematics. During his Italian journey he was thinking projectively, however, and penetrating to the archetypal nature of plants. It was a long struggle before he arrived at the 'archetypal plant' and was then able to 'invent infinite numbers of plants that must be consistent with this,' as he put it.[97] The discovery of the archetypal plant arose from a realization of the essential nature of plants that also involved visualization. A new dimension opened up for the awareness soul. This will be discussed in more detail in the final chapter.

Schiller had started to study philosophy and history in his late twenties. He was thirty-four when his efforts in these fields came to fruition in his *Letters on Aesthetic Education*. His struggles in the field of philosophy led to the perception of essential nature. At the same time awareness dawned of his life's mission; one can feel how he identified completely with his discovery and how this also proved to have general validity.

Further understanding of the awareness soul may be gained by considering the importance of the middle region in Schiller's case. The 'instinct for matter' arising from the lower pole provokes 'change', movement, 'sentient response'. The 'instinct for form' arising from the upper pole has shape and form as its goal and lifts human beings from the 'realm of phenomena' to 'unity of ideas'. One gives rise to thought, the other to affectivity. In the middle region the two instincts combine to become the 'play instinct' in which dead form and sentient life are united and 'living form', beauty, is achieved. 'Beauty uses the senses to guide human beings to form and to thought; beauty uses the human mind and spirit to lead human beings back to the sphere of matter and restore them to the world of the senses.'[98] Out of this middle region, which has to be re-created over and over again, human beings become artists in the most comprehensive sense; it is from here that the potential for freedom arises.

Many outstanding individuals produced their most important works in mid-life, as Schiller did, or laid the foundations for such works.

Dante was in his mid-thirties when he experienced his journey through the other world and his *Divina Commedia*. Luther was thirty-four when he nailed the theses that were to trigger the Reformation to the church door at Wittenberg. Beethoven composed his 'Eroica', *Fidelio* and Fifth Symphony, the very core of his artistic work when he was thirty-two, thirty-four and thirty-six years of age. Wagner was thirty-four when he conceived the general outline of the *Ring* cycle that was to be his greatest work. Napoleon was in his thirty-fifth year when he became emperor.[99]

Geniuses provide the best examples of developments that are less overt in most other people and are only in their initial stages during the mid-life seven-year period and the one that follows. The fact is that the awareness soul is only in the early stages of its development in humanity as a whole. Many aspects of awareness soul development are only evident in the more outstanding individuals whose creative work foreshadows the future.

The epoch of the awareness soul began with the present age, when humanity was taking new departures and–initially developing its new awareness at the level of the senses–entered the creative age of discovery and invention. It was only from this point forward that the awareness soul was able to develop in individuals who had reached the mid-life period and, as in the case of humanity as a whole, it first of all turned to the outer aspect of the world.[100]

As a result it has become possible for every individual to have greater awareness of their quest for the essence of things once the middle of life has been reached, and to live for the essence. One may try and develop presence of mind and spirit in the profoundest sense of the word, looking for the presence of the spirit even in everyday life wherever possible. In this sense every human individual can become creative once the middle of life has been reached. It may not be their destiny to make a major impact on the world, and they may not all be great artists. But everyone can be a master in the art of life in mid-life. Everyone can seek to recognize the 'idea', their special life mission, by considering the 'material' of life and letting it emerge–just as a sculptor uses his chisel to let the figure he has perceived emerge from the block of marble.

The objection that such a comparison rates individuality too highly, for after all, no one is irreplaceable, arises only from a failure to recognize the true nature of the ego incarnated in a human individual. The truth is that no one is replaceable.[6] Every individual has a particular mission, however small, whatever their position at work and within the family. Someone else may be able to do what they do or take over their social function; but only the particular individual is able to do it in that particular way.

Humanity is like a great painting; to make it a real work of art, not a single nuance of colour nor a single brush stroke can be omitted.

The vital question for the awareness soul of every individual during the sixth seven-year period therefore is: 'How do I find my way to the essence of the world and to my own true nature, and how can I bring my own true nature to realization in the world?'

Relationship to the Physical Body
The very genesis of the awareness soul has to do with this relationship. The ego has to go even deeper than before in transforming and refashioning what is there already. It now has to transform the powers of the physical body into the awareness soul.[78] The physical body is the part of the human being that has gone farthest into the solid state, and the ego therefore has to work much harder and develop maximum activity. We may visualize this as a precondition for increased ego activity in the awareness soul. In coming to grips with physical powers that have gone entirely into form, the ego also attains to the wide-awake state that is initially directed at the earth world, that is, the world of which the physical body is a part. A relationship may be perceived between the sixth and the first seven-year period, and we shall see this more clearly when we consider abnormal developments relating to them.

The relationship between the awareness soul and the human principle in which it has its roots continues until that life comes to an end; we have seen that the same holds true for the other soul elements. The mid-life period has special significance in this regard, however, and this is principally due to the deterioration of the physical body. It is in mid-life that degeneration, catabolism, begins to be in the ascendant over anabolism in the physical body. Catabolism has been present from the moment the first breath was taken, but until midlife it was made up for by anabolism, which was greater. Now anabolism is getting less and catabolism comes to the fore. This provides the basis on which the experience of one's own death arises in the soul.

A number of physical signs give evidence of this fact. The vital capacity of the lung, i.e. its ability to inhale vitalizing oxygen, reaches

a peak in the thirty-fifth year.[101] Recent investigations have shown
that it begins to decrease even before that, but it is only in the early
forties that the graph shows a relatively sharp drop. The liver, the
'central organ for anabolism' reaches its maximum weight between
thirty and forty, and its weight goes down by about fifty per cent
in the second half of life.[102] People grow shorter once they have
reached the middle of life, and recent investigations have shown a
slight reduction in height as early as between twenty-eight and thirty
years. Later on the spine will develop more of a curvature as the
intervertebral discs flatten out. Human beings bend down again
towards the earth from which their physical body had originally come
upright.[103]

The curve of biological development that may be drawn to illustrate
these facts begins at birth and reaches its zenith in mid-life. It begins
to flatten out in the late twenties, and at about thirty years of age
it begins to go down, finally ending with death. As the diagram shows,
another curve comes to meet it. This is the curve of the essential
soul and spirit. Its lowest point comes very close to the zenith of
the biological curve.

The diagram shows that in mid-life human beings have really found
themselves. With the development of an independent awareness soul
the essential soul and spirit has entered deeply into the earthly realm.
Before this, this entity entered into relationship with a life process
that was in the ascendant in the physical body (and thus with the
natural world and the cosmos). This is indicated by the arrows on
the left which go in the direction of anabolism and incarnation, which
predominate in the first half of life. They grow shorter as life progresses toward the mid-point and this represents the fact that from
the moment of birth anabolism gets less and less. (With these arrows
and the ones shown on the right we get a metamorphosis of the
diagram presented by Holtzapfel.[103] They serve to indicate the way
the living physical body is taken hold of and organized by soul and
spirit.) The higher birth processes through which catabolism and
the first stages of excarnation show themselves even in the first half
of life have not been included.

The awareness soul is born in mid-life, at a time when catabolism
and excarnation begin to predominate. Conscious awareness cannot
arise unless there is catabolism, as is evident from the dominance
of catabolic processes in the brain and nervous system. Every act
of awareness, every sensory perception can be seen to involve

degenerative processes. It is of paramount importance for the increased awareness of the awareness soul that it goes hand in hand with a predominance of degenerative processes. The living physical body withdraws before the clear light of that conscious awareness, liberating it as it undergoes degeneration. Its metamorphosed powers, however, ascend with the awareness soul to the realm of the spirit.

In mid-life the ego finds itself between spirit and body. The spirit is asking it to join it in the heights. The human being on the mountain top is looking up to the heights. Yet the earth, on which a firm foothold has been gained, also exerts a greater pull than ever, holding the human soul fast through senses that have come awake.[104] If individuals follow the spirit to the heights the zenith of the biological curve becomes a turning point in the development of soul and spirit that until the late twenties had been going hand in hand with biological development. Now it gradually joins the curve of the essential soul and spirit and having entered deeply into physical life and come close to the curve of physical powers as they reached their zenith, this now moves upwards again towards the realm of the spirit.

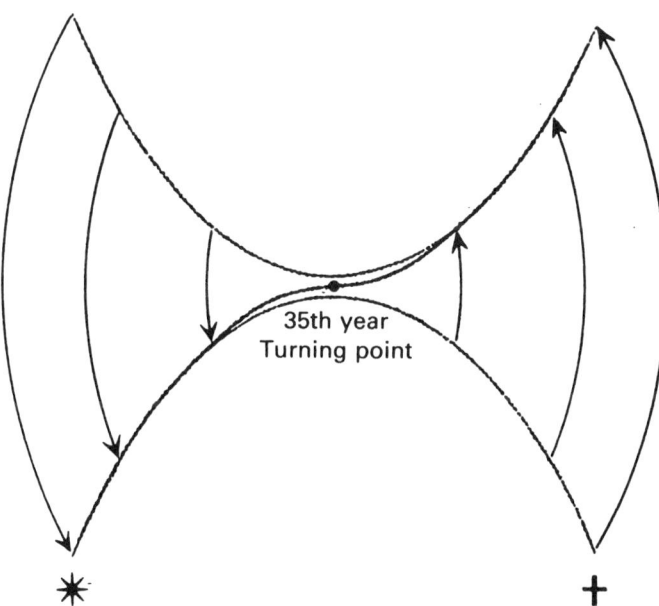

35th year
Turning point

The curve representing development of soul and spirit separated from the biological curve in about the twenty-eighth year, though at that stage both curves were still ascending. Inner soul development took physical consolidation as its base and at the same time began to rise above it. Then, in mid-life, the two curves come to move in different directions.

The biological curve moves down, and physical development goes downhill in all human beings, more in some than in others and more rapidly in some than in others. When the curve for the development of soul and spirit has gone through its turning point, the life of the mind and spirit can gain new impetus and move upwards. If however this turning to the spirit does not happen, or if it is incomplete, the curve of soul and spirit will more or less continue to follow the biological curve. It will deviate from the curve for the development of soul and spirit, and the essence of the human being cannot emerge in the course of further development. The life of the soul is drawn into the increasing degeneration of the living body. The mental disorders and diseases that arise in consequence of this will be discussed in a later chapter.

According to Hampe, the decision as to 'whether the rest shall be a withering or a ripening process' is made in mid-life.[105] This will be the first but not the only time that it is made, for the decision is made over and over again in the second half of life, or at least it should be. The arrows on the right may therefore serve to indicate that even if there has been a turning point in mid-life, it will be necessary to gain new impetus again and again in rising above the downward-dragging tendency of the physical body. If there has been no turning point, or if it has not been fully achieved–the latter probably being the case for the majority of people–it can be made up for by new impetus being gained later on. The arrows are getting longer as the end of life approaches and thus also represent progressive excarnation in the second half of life. Powers of soul and spirit are increasingly becoming independent of the degenerating physical element. The longest arrow on the left marks the moment of birth, the longest arrow on the right the moment of death.

With regard to the physical body, the question also arises as to which organic system provides the base for the awareness soul. The connection between conscious awareness and nervous system points to the brain; the gestures in which the awareness soul comes to expression suggest the head. Steiner drew attention to the habit of putting a finger to the nose.[6] Another gesture–the hand supporting the head–also fits the picture.

The awareness soul rises from the head to enter into a spiritual life of perceptive thought. We have seen that this form of thinking must include powers of will and feeling if it is to be alive. Steiner spoke of 'reverent devotion' as the educator of the awareness soul.[106] In reverent devotion the will is given up to higher things and fills with love for them. Reverent devotion also has to do with thoughts and with concentration. Like love, reverent devotion means that soul forces from below, from the middle and from above are working together. And as we have seen, the heart supports such concerted action; we also have direct experience of the connection between the heart and reverent devotion.

With the heart we have come back to the 'centre of the soul from where access can be gained to the true self'.[106] There the encounter takes place between lower and higher ego that is most characteristic of the awareness soul. The awareness soul lives out of the head, but draws its life from the heart. The thinking in our heads would die could it not be given sustenance by the heart. The light of thought becomes the cold light of the intellect that illumines only the surface features of the earth unless it takes in the warmth of heart that goes deeper and conveys will, love, and enlightenment. Without the will that moves and without love for the object, no line of thought can be followed that will give enlightenment. When will and love unite in reverent devotion the thinking of the awareness soul opens up to the higher life.

Summing it up in a Picture

The following picture may serve to summarize the development of the different soul principles. The sentient soul may be compared to a plant in its leafy stage. A plant opens up to the world through its leaves; in the same way soul life opens up to the world through the sentient soul, and like a plant it is built up in the process. The intellectual or mind soul as it grows more inward corresponds to the development of flower bud and calyx. The awareness soul, which opens itself up to the spirit and out of which the spirit is acting in the world, may be compared to a bright flower that reflects the light of the sun. In the second half of life, the flowering of the awareness soul will become a fruiting process and human beings will be able to give back to the world part of what they have received in the first half of life.

This picture can also help us when we consider the question as to what happens to the other two soul elements once the awareness soul has 'come into flower'. One often hears it said that surely we then no longer have need of a sentient soul. That, however, would be just like saying that once the plant has produced a flower it no longer has need of leaves. But if we were to think that and take the leaves off a plant it would soon wither. Something very much like this would happen in the life of the soul if it were unable to draw constant new life from the powers of the sentient soul. The sentient soul–and correspondingly the intellectual or mind soul–can be a source of renewal that will give new strength throughout life. We shall return to this in the discussion of developmental disorders of the awareness soul and in the last chapter.

If the above renewal of life is to be healthy, the sentient soul and the soul elements generally have to be transformed so that they are always in accord with the stage of life that has been reached. Distinction must be made between individuals whose soul elements undergo transformation to provide for progressive development, and individuals who want to keep a soul element in its original form. Soul development is not helped by holding on to a particular stage in life, nor by skipping any stage, for its own time form is also progressively changing.

In about the forty-second year the three seven-year periods of soul development come to an end, and the spiritual development of the individual makes itself felt in the biography–or at least should make itself felt.[107] Soul development being the subject of this book, the description of biography will not be taken beyond this point.

Chapter Three
Developmental Disorders of Soul Life

Developmental Disorders of the Sentient Body

The major disorders of soul development and the psychological conditions resulting from them are due to problems arising during the six seven-year periods that go up to the early forties. After that one merely sees age-induced metamorphoses of the basic disorder.

Developmental disorders relating to the sentient body show the characteristic polarity of the sentient body itself. This may be weak and underdeveloped, or it may grow too powerful and exceed its boundaries.

If the sentient body is underdeveloped, the impression is one of more or less marked infantility showing itself from puberty onwards. Young people appear childlike in a way that belongs to the second or even the first seven-year period.[1] They like to take the lead from others, either by imitating people the way children do during the first seven-year period, or by depending on, and indeed clinging to, someone in authority, a trait that may persist for the rest of their lives. Independence of parental authority is not or only partly achieved in puberty. Considering their soul life, one still has the impression of it being 'cocooned' in the way that is characteristic of pre-puberty.

This may affect the whole of soul life, so that the intelligence is also underdeveloped. The result is a greater or lesser degree of mental deficiency which may or may not involve diagnosable brain damage. In the majority of cases, however, only the emotional life and the will are affected, and 'intellectual development is appropriate for the age'.[2] Emotional reactions seem childlike, with feeble emotions, sometimes also rebellious reactions; the will is always weak and there is a tendency to sulk childishly. Others may show a childlike carefree attitude and none of the above reactions. Infantile young people want to hold on to the security of childhood; all their life they will keep returning to a 'zone of safety' where they are able to 'enjoy life in play'.[3]

They also show signs of physical underdevelopment of the kind described by Kretschmer.[4] Such underdevelopment may involve the organs of reproduction and sexuality and also the rest of the physique. On the other hand physical and sexual development may be normal, and indeed the latter may be excessive, but emotional development has not kept pace. In this case the sexuality may have compulsive character. In anthroposophical terms this means that the astral body was able to achieve development and maturing processes up to puberty but was then unable to separate from them properly. Its birth from the physical body is feeble, but this should not mean that the astral body as such is feeble.

Overweening vitality of the sentient body may be due to a physical or emotional infantilism the core of which has not been overcome. The symptoms of puberty that have been described above become excessive, with exaggeration of opposite extremes. The tendency to criticize becomes constant criticism and a totally negative attitude, and this may be combined with a marked tendency to think in abstract terms. Emotional reactions exceed the boundaries and lead to aggression and even destructiveness. Sexual desire is enhanced but does not go hand in hand with the power of love. In many cases the power of love is there, or at least a longing for it exists, but it is still at the level of childish adoration; sexual drives seek their physical satisfaction independent of it. One can sense an inner uncertainty and weakness in these young people, a sign that the emotional life has not sufficiently matured and is being carried along by the powers of puberty that have now come to life. It is also possible to sense a desire to get on top of their painful inner infantilism by being excessively active in the sentient body.

Activity may thus be due to deficiency, but it may also arise from superabundance, in which case it presents itself as a more elemental force. This type of excess, which is not due to infantilism, will lead to a new kind of retarded development if it continues beyond the twentieth year. Infantile dependence on parental authority may persist throughout life and so may exaggerated protest against parental authority, which may even prevent the individual from establishing a new family unit.[5] The social problems that arise differ from those seen with underdevelopment. In a young person who has remained infantile, problems arise because of passivity, a marked tendency to follow the lead of others and depend on them. An excess of soul powers on the other hand may lead to criminal acts. Having separated from their parents for good and left the family home, these young people often begin to roam, start thieving and become delinquent. Infantile young people may take this route through being misled and dependent.

Autonomic Disorders
Specific neuroses arising from the developmental disorders described above will not be discussed here, but some consideration must be given to the autonomic disorders that are getting increasingly common in young people, for these may inhibit soul development for the rest of an individual's life.

Again two opposing trends are apparent, and these may arise through weakness or strength in the life of the astral body. It has already been said that the astral body relates to the metabolic sphere in the organism through the urogenital system. Following its birth from this sphere it continues to send its rays into the organism via the urogenital system and is responsible for involuntary movements in metabolism and in emotional life.

Weakness in the astral body means that the rays sent through the organism are weak or, in short, that kidney radiation is weak.[6] The astral body does not provide sufficient impetus for the metabolism and the blood circulation which is maintained by metabolism, so that they become sluggish and lack tone. The result is low blood pressure-always an indication that astralization is inadequate-and poor peripheral circulation. This is predominantly the asthenic constitution, and as the name suggests, this is marked by weakness. Mentally one also sees a lack of tone, and this takes the form of lameness or of flagging early, rapid mental exhaustion, or a weak (non-hysterical) outflowing into the world.

Strong kidney radiation represents excessive astral activity, resulting in great metabolic activity and raised blood pressure with a tendency to powerful pulsations, flushes and other circulatory processes that are enhanced or of which there is enhanced awareness. This is predominantly the pyknic type with the emphasis on metabolism.

Mentally one notes a powerful, uncontrolled emotional life that may build up and then explode in fury and aggression.

Both these dynamic constitutional types show physical and mental spasms. With the weak kidney radiation of the first type they signify deficiency and the reaction of a weak astral body that is trying to intervene more strongly in the physical sphere but gets stuck in spasms and temporary rises in blood pressure. With the strong kidney radiation of the second type spasms develop because there is superabundance; the astral body is losing its hold in the physical body and is seeking to take hold in spasm. That is also the basis for the two common types of headache and even migraine that need to be treated differently depending on whether the patient belongs to one type or the other.

These and other phenomena clearly show the polar extremes in the astral body. Here one gets a first overview over the vast array of autonomic and vasomotor disorders. Even in young people these may in extreme cases take the form of acute nervousness and angioneuropathy. Guidelines also emerge for medical treatment that is not symptomatic but goes to the root cause.

It is also possible to relate the two groups to the developmental disorders we have discussed. Experience has shown that infantile young people with underdeveloped sentient bodies are more liable to have weak kidney radiation and the symptoms arising from this. Constitutionally they tend more towards the asthenic type.[1] Young people with an overweening sentient body generally show the signs of excessive kidney radiation. Transition and mixed forms also exist and are found in adults as well as in young people. It is also possible for the mental effects of a physically weak or excessively strong kidney radiation to be overcome by the ego. Resistance to the basic constitution may help individuals to develop considerable powers that help them to overcome and grow free from constitutional traits, and this may also lead to improvement in physical symptoms. A young person whose ego is not yet born will however hardly be capable of this and therefore needs special help also with autonomic disorders.

Disorders of Basic Soul Elements
The power of desire is reduced in infantile young people; it is enhanced in those whose astral forces are excessive. In both cases the power of discernment by which issues are judged and transformed

is not sufficiently developed. This is immediately obvious in infantile young people, particularly if they are mentally defective. The other type may well show a strong inclination to form opinions, but this does not derive from an ego that seeks to understand but from excessive astral activity. Opinions are not merely prompted and guided by sympathy or antipathy but entirely determined by them, so that they remain entirely subjective. It is not opinions so much as prejudices that are formed, or people and issues are condemned without proper judgement.

Sentience–and judgement plays a role in its development–is also underdeveloped in both cases. Over and over again one hears young people from both groups say that they cannot feel things properly, and they often complain of inner emptiness. At the heart of this lies an emptiness in the sentient sphere. This may combine with a feeling of being frozen, and indeed dead, inside and may drive some to the point of suicide. Steiner said in 1910 that 'deadly boredom' was a pathological factor;[7] today it has again become a major threat.

Young people like these may be presented with the most glorious sights and sounds–provided they even want such things–and they will feel nothing or hardly anything at all. They may travel to marvellous places but the response is nil. One can feel that there is a desire for beauty wanting to arise deep down in the soul, but this clearly does not connect with the sensory impressions they receive, leaving them 'cold'. The union of sensory perception and inner desire that gives rise to sentience does not occur. Initially this is also the reason why some young people who still do feel something lack differentiation in their sentient life. This is evident from monotony of verbal expression. Totally different things that make an impression get the same response, 'fab' perhaps, or 'great', and there is no faculty for greater differentiation in their appreciation.

We have seen that the union of desire and perception can only occur if desire relates to the world in greater awareness, that is, if it is combined with perceptive discernment and is transformed into an interest that is alive and active. Seeing the pathological side of this, it is even easier to appreciate that this is the only way in which impressions can be properly taken in and digested so that a differentiated response may arise in the soul. Once again the blood may serve as an illustrative example. In so far as it relates to the food we eat, blood can only be produced if that food is properly assimilated. In the same way, sentient responses can only develop if sensory impressions are at least to some extent 'assimilated'. Failing this, mind and soul will suffer from malnutrition, however rich the material offered by the world may be, and the sentient body and later the sentient soul become 'anaemic'.

This situation may form the basis of addiction and it does so much more frequently than is generally thought.

Addiction as a Form of Developmental Disorder

Addiction always means that a desire felt by the astral body has become a compulsive habit. Hunger and thirst are habitual desires that have to be met. Compared to those physical desires addiction is primarily a desire in the soul; it is just as compelling in having to be met, however, as the thirst felt when one may be dying of thirst. Just as in the latter case, an addictive soul's desire becomes instinctive. Habit and instinct, however, serve to maintain life and are rooted in the life body,[8] which establishes the transition from soul to physical body. One can experience how in addiction desire goes in the direction of the physical body. The more intensely it connects with the life of that body, with its life body, the greater and more vital is the physical body's need to satisfy what has now become a physical desire.

This is particularly evident in drug addiction which is becoming more and more of a burning problem in the third and fourth, and indeed even in the second seven-year period. The incidence increases up to the early thirties, with twice as many males involved than female. Here, too, the female sex remains in a healthier state. The histories of drug addicts show a high incidence of mental and emotional instability, inner uncertainty, infantile and depressive traits. A tendency to dissocial behaviour comes to expression in vagrancy, thefts etc. It is enhanced under the influence of drugs.[9]

In the context of the developmental disorder under discussion, addiction may develop if there is underdevelopment, and also if there is excessive activity in the sentient body. When infantile young people with their feeble emotional reactions face the crisis of puberty they may develop the urge to look for the powerful inner experiences that can be had under the influence of drugs. The desire to join one's peers or be subject to them is a further impulse supported by the shared life of the group. Drug consumption may be an 'initiation' test and lead to addiction. On the other hand childish curiosity may be the trigger.

Excessive vitality in the sentient body produces a chaos of inner forces that the young person is unable to cope with and seeks to escape by taking drugs. This may take the form of conflict situations or of protest against parents, school or the establishment. The fundamental element is always the inner state that has been described above, and this is connected with the developmental disorder.

This state of mind, which we have already discussed in considering the developmental disorder as such, is very evident in addicts. We have seen how opposite extremes become exaggerated in soul life and how this was also apparent from too few or excessive soul forces. No real effort is made to establish a centre that would bring order and gradually control the emotional instability. One will find that the centre has been largely lost, a term used in the history of art that may also be used in psychology and psychiatry.[10] This tends to be more marked in males than in females.

Between fourteen and twenty-eight years of age, sentience lives out of the centre of a soul that looks toward the world. It is the lifeblood of a young soul that today threatens to grow anaemic. The histories of drug addicts very often mention the inner emptiness referred to above,[11] at the core of which lies a weakness of sentient life. The fact that boredom is regarded as one of the factors in drug abuse[9] indicates that there must be a second component to it.

Boredom is not merely emptiness, it is also a 'hunger' of the soul, as Novalis put it.[12] This hunger cannot be satisfied by what the world normally has to offer, for these impressions are not assimilated. The soul therefore looks for other stimuli that will quickly and without need for effort fill up the emptiness inside and bring movement into the frozen inner life. Such stimuli are provided under the influence of drugs. In the majority of cases the experiences are only short-lived, however, leaving an even more painful void once the drug has ceased to act.

The most familiar example of this is the 'hangover' that follows the euphoria of alcoholic intoxication. Chronic alcoholism, the commonest form of addiction among adults, is less widespread among the young, though it is steadily on the increase among them. Young people first show dependence on alcohol in the form of 'leisure time alcoholism', which is rampant in that age group. Alcohol consumption frequently takes the place of other activities such as dancing and conversation in that situation and is used as a drug to overcome the experience of emptiness in leisure periods.[9]

With alcoholism, too, desire increasingly becomes a physical craving until the alcoholic is no longer able to exist without drink. The ego comes to be completely submerged in the soul's overwhelming desire, and the body suffers damage due to alcohol as well as burning desire, damage that affects particularly the liver as the metabolic organ for ego and ether body. With alcoholic intoxication, 'indulgence' extends to the inner organism, with the result that its sensory perception become too intense. The tasting function of the liver in particular grows too conscious and interferes with the metabolic processes that are at an unconscious level.[13]

Drug Addiction

Drug addiction in a wider sense may be due to the abuse of hypnotics and sedatives as well as of stimulant drugs and indeed certain psychotropic drugs. After some time people are no longer able to do without the sedative, tranquillizing, stimulant and sometimes even narcotic effect. The widespread dependence on tablets one sees today leads to mental and physical changes. It may trigger drug addiction in the narrower sense, i.e. addiction to narcotics.

The major drugs used by young addicts are hashish ('hash'), the resin of Indian hemp; marihuana, the dried flowering and fruiting tops of the same plant; LSD, which is derived from ergot, and mescaline, from peyotl cactus. Hashish smokers are still looking for the kind of euphoria that is also provided by alcohol and cocaine, but visionary experiences also develop and for this reason the drug is sometimes called the 'leaf of delusion'. The stronger hallucinogens LSD and mescaline go beyond fabulous and fantastic visions and dissolve the sense of position in space.[14] Colours separate from objects, shapes dissolve in movement, a 'chaotic kaleidoscope of form and colour' develops and combines with the sensation of 'spreading out into the cosmos', with the ego dissolving in cosmic harmonies. Then again a flood of new images in intense colours condenses itself. Such a psychedelic (i.e. soul and spirit revealing) trance involves experiences of 'death and eternity', the addict feels transported into supernatural worlds.[15]

Judging by the reports one hears and reads, the actual trance state is followed by powerful and indeed ecstatic sensations. But again these experiences and sensations do not help the soul to develop; instead, the desire arises to repeat the 'trip' and escape from a world that is grey and rigid. 'Sober observers have noted that the religious ecstasy does not persist; such experiences melt away again and do not become a focus or motive for action.'[16] The psychological reason for this is that judgement, the other central activity of a young person's soul life, is largely eliminated. Desire is directed immediately to the narcotic experience; it is not induced to assimilate the experiences and gain insight. The hot desire of the astral body is not transformed into a warm interest on the part of the ego that would make those experiences more lasting.

Physiological cravings at the physical level are not normally seen with these drugs. There is however marked paralysis of the will after repeated drug trips,[14] and this makes drug dependent individuals completely unable to cope with life on earth. The physical basis for this paralysis of the will and for other psychological symptoms emerges from the fact that LSD is predominantly stored in the liver and kidneys.[15] This brings to mind the connection between the liver, will activity and depressive states—these are commonly seen in drug addicts—and the 'lethargy' of emotional life that has to do with the renal system, an aftereffect that has also been mentioned by a protagonist of hashish.[14]

Ernst Juenger said: 'The danger is inherent in the individual, not in the thing itself, and any particular inclination may therefore lead to addiction.' This is undoubtedly true. Television, devotion to sport or the sexual experience can all lead to addiction, as can the desire to destroy, even if it is initially based on philosophical or political convictions. Strong, mature individuals will certainly have a chance of coping with drug addiction by themselves and gain temporary stimuli for creative work from the use of drugs. Without going into the problems that exist for adults—the words 'temporary' and 'stimuli' give an indication—Suchantke's cogent argument against drugs for young people is worth quoting: Young people have 'not yet developed mature personalities' and therefore run much greater risk of becoming addicted to drugs.[14]

It may be maintained that smoking hashish is not addictive and does not cause personality changes if kept within limits. Yet there always is the possibility of its reducing the threshold of inhibition from using harder drugs in unstable individuals. That is how hashish smokers become 'fixers' and soon find it impossible to manage without their shot of heroin. Yet heroin, a morphine derivative, is the most powerful narcotic known today. It will quickly banish anxieties and enhance the mood, but it wreaks destruction even on the physical body. The infantile helplessness of a heroin addict who is wasting away and is desperately pleading for a fix when almost at death's door reveals the extent of regression into a pathological childhood state.

Longing and Addiction

The experiences reported under the influence of hallucinogens reveal another motivation that plays a role in drug addiction. A report on the 'miracle drug' LSD says: 'There is a religious and even metaphysical hunger in the souls of the young that simply cannot be satisfied by traditional religions' (Allan Watts, quoted in [17]). This hunger drives many young people to the miracle drug which they hope will open up the supernatural and divine world for them.

It is a world they want to experience above all in images, as Buehler also stresses. The metaphysical hunger arising from emptiness of soul is primarily a hunger for vision. Bored with television and the cinema, the young people pounce on the world of cosmic images that arise under the influence of drugs. They are in fact longing to go back to a level of awareness that existed in earlier times when humans received revelation in living images from the world of the spirit.[17] The hallucinatory experiences of drug addicts arise from a pathological recrudescence of that old visual awareness.

Steiner gave a course on the education of the young in 1922 in which he spoke prophetically of future generations of young people who would no longer feel the enthusiasm of preceding generations but a much more powerful and indefinable longing, nay yearning.'[18] This indefinable longing has a core to it rather than a goal. At its core it is a compelling longing for a supernatural world of the spirit. When it becomes a bit more conscious this urge leads young people to join youthful sects. There the things of the spirit are not conveyed in the kind of training of awareness that would meet the needs of the present time; instead an intoxicating atmosphere is produced by more or less subtle means and the ego is carefully and deliberately left out of play. It has to be said that this has helped quite a number of drug addicts to overcome their addiction. In terms of progressive development to attain to the spirit it must however rank as retrogressive.

An increasing urge to use drugs to obtain surrogate gratification of the soul's or the body's desires indicates two things. In the first place it has to be admitted that compared to the early part of this century the longings of young people are generally more indefinable and hidden. In some cases it is merely a longing to have longings. But this more deep-rooted longing is also stronger and more elementary. Once the most profound longing has risen from the soul it turns as an absolute matter of course and without hesitation to the spirit, much more so than was formerly the case. The question arises as to how this longing can be released from the depth of the soul and brought to awareness.

Seen in the context of biography, the desire for the spirit in the souls of young people is the longing for individual soul development. The sentient life of the soul wants to unite in a living way with the spirit which it is seeking to attain; sometimes the inner emptiness can be experienced as a silent plea for this. If this dark power of desire is not raised to the light of conscious awareness longing may become addiction. The longing soul is seeking a goal that will bring fulfilment. If the longing is drowned in desire, the seeking longing turns into addictive desire that finally becomes physical. It becomes a sickness in its seeking. (The two words have the same root.)

Supernatural Experiences due to Drugs?
Drug addicts will say that it is exactly in the ecstasy of a psychedelic trance that they leave their bodies and enter into experiences comparable to the revelations that have come to great religious figures and artists. Surely narcotics were formerly used for religious rituals in some tribes?[11]

Only brief reference can be made to the mode of action of such drugs. In every case vital ecstasy, i.e. a loosening of powers belonging to the ether body, arises in the metabolic sphere, and particularly the liver and kidneys. These powers bring their own intense life into the sphere of the senses. That explains the newness, vitality and intensity of the perceptions. The individual is also 'beside himself' where his soul is concerned; that is evident from the floating sensations and the expansion into the whole cosmos that are experienced. The condition may be compared with a 'tremendously enhanced dream experience',[15] where the soul is outside the body. Steiner has also spoken of this with reference to people under the influence of opium.[20]

Compared to the effect of opium, when there is a definite reduction in conscious awareness, conscious awareness is not reduced but certainly changed by the hallucinogens LSD and mescaline. The situation is similar to the one seen in schizophreniform psychosis. This will be discussed later, but it may already be said that it involves a partial death process; not only does the astral body leave the physical body, the way it does in sleep, but powers of the ether body do so as well, transferring their life to the soul sphere. LSD intoxication produces an analogous picture. As with many schizophreniform psychoses, conscious awareness is not reduced but rather extended, made to come alive and intensified. It may be deduced that the astral body and ether body, both in the trance state of separation from the physical, maintain intense connections with the neurosensory system.

It should be noted, however, that this refers to individual soul faculties; the wholeness of soul life is lost in a kind of schism. There are occasions when individuals under the influence of drugs observe themselves from somewhere outside.[21] This brings intoxications of this type very close to schizophreniform psychoses. Hallucinogens truly live up to their other name of 'psycholytic drugs'. Mescaline in particular produces symptoms very similar to schizophrenic disease and the hallucinations that go with this; it may even trigger a genuine schizophreniform psychosis. (For hallucinations, see the chapter on schizophrenia.)

It is true, according to Steiner, that any form of drug intoxication that involves an opium-type trance and results in a 'trip' to other worlds gives certain insights into the world of the spirit.[20] Under the influence of these drugs and in psychosis the soul not only separates from the body but also enters into non-physical worlds. It is however pursued by powers of the individual's own metabolism that have been mobilized by the drug and enter into the soul. The supernatural experiences of drug addicts are thus clouded, distorted and given physical connotations by those metabolic powers, to the effect that the character of those supernatural worlds is changed and appears physical.

Spiritual training leads to experiences that are merely comparable to colour visions;[22] drug-induced hallucinations greatly intensify the colour experience of the senses. In this case the situation is not actively controlled by the ego the way it is with spiritual training. Drug addicts give themselves up passively to the effects produced by a drug that forces them out of the body without letting them achieve real freedom. Once the euphoric state has passed they are more than ever caught up in the body, and in their lethargy and depression they experience it as a prison; its sufferings, which have been induced by the soul, in turn cause further pain to the soul.

There were earlier times in human evolution when a relationship to the body had to be struggled for and won. Then there would have been a point in using narcotics to connect religious experience, which was in images, with the body and its powers. Today the connection with the body has been made and narcotics no longer serve a true religious or philosophical purpose.[19]

The Origins of Developmental Disorders and Addiction
It is easy to see that developmental disorders in the third seven-year period have their root in the preceding seven-year periods. Those two periods also hold the key to all subsequent disorders, and we shall therefore consider some of their pathology.

The histories of patients with developmental disorders and of drug addicts show a certain hereditary element and a high incidence of family and school trauma. One element is lack of contact with parents who in turn do not give enough of themselves to their children. There are other cases where the children are given too much attention and thoroughly spoiled. Sometimes the parents themselves show pathological instability; they fight a lot and are habitual tablet-takers, really setting an example for the future addict. Many parents are caught up in material things and it is their aim in life to be comfortably off. This is yet another element, and young people often also use drugs as a form of protest against such attitudes. Others use them to help them escape. They may have wishful ideas of another, better world into which they can escape, and drugs provide a pseudo-fulfilment of such dreams.

The family situation has been thoroughly investigated, but little mention is generally made of the role played by school life. This does not mean, of course, that school does not have an effect, but rather that the effect is so widespread that it does not show specifically in individual cases. Lutz quite rightly calls 'school education that is largely intellectual and geared to achievement' a factor to be taken into account when treating drug addiction. He also states that this type of education can cause 'a discrepancy to arise between goals set and achieved', with relatively too much demanded of the children, and this helps to create the situation in which drug addiction arises.[11]

The above and other factors finally lead to the developmental disorders that have already been described. In the final instance those disorders are signs that the incarnation of the essential child is impeded. We have seen that where the astral body is concerned this incarnation takes place until maturity for life on earth is achieved in puberty. It is only at this point in time that the soul has reached the lower physical body and hence the earth; the ego waits until it has achieved independence before it comes down to earth. The ether body moves in the opposite direction and out of the physical body. Around the seventh year, when soul and ego are about to penetrate the physical organisation from the head region, the ether body progressively separates from the physical body in the region of the head.

The above family and school traumas all interfere with that counterpoint of movement in the developmental process. The movement of the incarnating soul and ego is impeded, and the movement of the excarnating ether body is pathologically enhanced.

Family
In family life the process starts in the first seven-year period. This is the period when children are in empathy with the world around them and imitate it, thus creating the physical preconditions for the incarnation of soul and ego, their 'attunement' to the body. If parental contact is poor, empathy and the imitative behaviour that plays a role in organic development suffer. A precondition for this imitation is warmth, which at this stage still relates intensely to the form-giving nerve functions of the head. The development of body warmth in children partly depends on human warmth. The warm nest of family life is partly converted to physical warmth by the child; it not only creates emotional contact but also 'incubates' the physical body of the child.[23] If there is not enough of it, the ego, which lives in states of warmth and unfolds its creative activity on this basis, cannot fully in-form the body.

Imitative empathy is also endangered when the person to whom the child relates will not let go and the 'nest' becomes a hothouse. Soul and ego will then withdraw into emotional coldness and this may lay the foundations for excessive opposition and desire to criticize later on. Parents tied to material values and always fighting each other also have such a negative effect on their children that these withdraw. In other cases parental instability is transferred to the child through empathy and prevents the healthy incarnation of soul and ego.

In all these cases, soul and ego remain too much in the head after the seventh year, having been prevented from taking the path that leads to the body. Education with the emphasis on intellectual training has the same damaging effect.

School

As De Rudder said in 1959, 'School children are faced with a volume of knowledge that is of sheer infinite proportions in the different subjects.'[24] Today more than ever all that can be done is more or less to cram those facts into the children's heads. This results in the much-quoted 'pressure to achieve' that becomes the 'pressure of pain'. Excessive demands on the head mean that soul and ego are onesidedly concentrated on this region, and pressure to achieve also means that the soul cannot find the strength it needs to enter into the body. This is where authority wielded with love should give guidance. A teacher who has to convey masses of information is only able to do this to a limited extent, if at all. Instead of emulation in accord with individual gifts the result is infantile subjection or intellectual precocity.

In addition cramming tends to go hand in hand with a tendency to insist on one's own opinions at much too early an age, that is between the ages of seven and fourteen; children will therefore either escape into infantilism and the imitative stage of early childhood,[25] or reproduce the opinions and ideas of adults at an intellectual level. Infantile children parrot others, precocious, intellectual school children reproduce what their minds have been trained to memorize.

A truly human education however means that the whole human being must be addressed in teaching, i.e. not only the faculty of thought, but the life of feeling and the will, i.e. the rhythmical system of the body and limbs and metabolism. This is now realized to some extent. It is the only basis on which an individual judgement that also involves will and feelings can develop. Then the young can form their own concepts that are based on discernment, and the conceptual organs of learning develop. Young people are prevented from digesting perceptions and experiences not only because their interest has not been sufficiently aroused, but because they lack such organs. Unless one has taken sufficient interest to learn at least something about Greek civilization, one usually has no real idea as to what to do with the impressions gained on a trip to Greece; the 'organ' for this is lacking. It may of course also happen that such a trip first of all arouses interest in the subject, and the learning process may come after the trip.

The truly human education that has been the aim of Rudolf Steiner's work in the educational field has formative influences not only on the soul but also on the body. It helps the body to retain the creative powers it needs in later life, powers that are also needed for the physical basis of feeling and will activity, and it also gives new life to those powers. One-sided emphasis on the excarnation of these powers and excessive involvement of those creative powers in the activities of the head causes the above-mentioned interference with the counter-point in developmental movements. Fixation of soul and ego in the head goes hand in hand with extremes in the metamorphosis of formative powers in the head region.

Any fixation of soul and ego in the head in itself means underdevelopment of the emotional and will faculties, and at the same time intellectual faculties are developed too early and too powerfully. The retardation in character development that arises is combined with acceleration in the intellectual sphere. The latter always means premature specialization and 'loss of creative potential'.[26] Excessive conversion of creative powers into powers of thought affects not only the life of the physical body, which does not come to full maturation, but also the life of thought. Thinking loses flexibility and adaptability whilst the tendency to abstract thinking that develops with puberty becomes excessive. A further precondition is established for premature and intense death processes in the thought life of the head; during and after the third seven-year period those thinking processes will lack the power to use judgement to control desire and let it develop into sentience and perception.

It may also be said that the 'indefinable longing' that is growing more and more powerful today and may indeed show itself quite early on in life is not taken into account and satisfied by educationalists. The abstract concepts taught at school become fixed in the souls of young people and present a further obstacle when they seek to fulfil the longing for the spirit later on in life.[27] (For further details, particularly with regard to physical effects, see the chapter on schizophrenia.)

Emotional and Organic Rebellion
When young people are infantile and their astral body is underdeveloped this astral body has not been able to incarnate fully. An astral body showing excessive development or on the way to having such excessive development is however also due to incomplete incarnation. The weakness of will, judgement and sentient life we have spoken of indicates that–in conjunction with the obstacles to physical development–the astral body is not fully matured; it has not been

able to enter fully into the metabolic and rhythmical systems and undergoes pathological separation from them as it grows excessive. It is at the same time subject to its inherent polarity and this causes emotional rebellion to arise from the lower pole.

In puberty young people with their overweening soul powers will instinctively turn against the things that happened or failed to happen in the preceding seven-year periods. They have experienced the 'establishment' in their own bodies. The emotional rebelliousness of young people in puberty is thus not only ascribable to the tendency to polarize that comes at this time; the educational efforts of those around them carry a considerable degree of responsibility. Such rebellion can and should therefore shake up those around them. For the young people themselves it will however prove unproductive, like any rebellion, so long as it remains at this particular stage and if the longing for the spirit that lies behind it cannot be brought to birth.

The rebellion need not be limited to the emotions. An organic rebellion will also be noted, in the first place among drug addicts. The powers of the ether body that have been withdrawn from the physical body and wrongly made to serve the head now come into their own when the soul is loosened under the influence of drugs. They continue to move out of the body, keeping to the direction that has been imposed on them, and in the final instance seek their cosmic home in the drug experience. Here we see a further element that leads young people to addiction–and indeed to seek death. Many are actually aware that the addiction arises from deep down in life and that they are seeking their true home, of which this life has once been part, in their addiction.

Therapeutic Aspects

Below, the description of a particular disorder will in each case be followed by suggestions for development-orientated treatment. There may be a certain overlap between seven-year periods. The emphasis will be on treating the soul and spirit; medical treatment and curative eurythmy will only be mentioned briefly. (This is discussed in the chapter on psychiatry in Husemann/Wolff *The Anthroposophical Approach to Medicine*, vol.3.[28]) The therapeutic measures suggested for body and soul may also encourage those who are not medically qualified to help others and also themselves. In more serious cases it will however be necessary to seek medical advice. The chapter will not deal with individual case histories, and a general therapeutic approach will be outlined that has proved useful in practice.

Before one can start to treat anyone, the diagnosis must be made. This is a rule that always applies, though we are not referring to a standard form of diagnosis, or a label used to cover up the individual nature of the case. Using the anthroposophical approach to medicine, the diagnosis should include the whole human being and his or her disease. The first step therefore is to observe and examine the patient as accurately and sensitively as possible; then a first evaluation is made against the background of the human being, the essential principles that constitute the human being, and the process of development.

This kind of diagnosis takes account of the situation that exists for each of the constituent principles, and one will find that the pathology already contains the therapy which should be 'made to emerge from it', as Steiner put it. The therapy arises out of the *essentia*, the true nature of the disease; it becomes the 'essence' that will heal the disease. One also finds that every individual has his own disease but that on the other hand every disease is an objective entity that presents itself against the background of human evolution.[29]

When it comes to treatment, the developmental disorders seen between the ages of fourteen and twenty-one (and later developmental disorders) first of all refer back to the first two seven-year periods. Once again one comes to realize that prevention is the best form of treatment. The developmental disorders described above are so common nowadays that treatment alone is not enough to cope with them. Families and teachers must develop an awareness for everything that will have to be done to prevent these disorders and for everything that should not be done. If this does not happen many children with less marked developmental disorders will go untreated and some of those milder forms will become more serious.

In the years between fourteen and twenty-one, prevention and treatment consists in an education that develops the ability to judge, awakens an interest in the things of the mind and spirit and thus proves truly formative. We can base ourselves on the fact that everybody has the right to education and training until the age of eighteen. An outstanding example is the Hibernia School, a Waldorf School that has twelve classes and offers training in various trades.[30] The omissions of the first two seven-year periods also provide the main guideline for treatment. The principle is to allow the young people to catch up, the goal is to let them mature now.

Catching up and Maturing

As a first step, the hospital or consulting room should allow the young people to experience some of the human warmth that has either been lacking or taken pathological forms at home. The aim should be to let them develop a healthy emotional empathy, particularly if they have remained infantile or have relapsed into infantile helplessness. The authority aspect will soon come in as well. Many who have not been able to enter fully into the life of the third seven-year period are at heart still looking for the authority wielded with love that they did not have in the second seven-year period. This may certainly be masked by protest and opposition. Parents will also be able to make up for some things if they are given counselling or indeed treatment, which will be necessary in some cases. In any case they can learn to have more understanding for the situation their children have reached.

As for school, they can be helped to cope better with problems at school by an approach based on the study of man and also on the medical aspect. This opens up another aspect of catching up for the therapist. Young people with developmental disorders have missed out on proper education and therefore also need to catch up in this area. It may be helpful to suggest private coaching. The aim of such coaching however should not be to provide the factual knowledge poor students have failed to gather, but to awaken their interest in such knowledge.

If such coaching is not available or if it is not enough, the therapist will have to take an active part. Physicians can never take the place of educators, but in cases like these treatment will have to include an educational element, just as on the other hand education should today include a therapeutic element.[31] Curative education, a discipline that is chiefly devoted to developmental disorders and mental illness in the first two seven-year periods, combines healing and teaching to the highest degree. It is however also possible to have curative education in a wider sense, and this should play a role in the treatment of later developmental disorders and mental illness.

In these cases, part of the individual therapy sessions in hospital or in the consulting room is devoted to a limited educational programme. One may for example let these patients report on something they have been encouraged to read; or describe anything they may have observed in nature the day before, as this may encourage them to go further into the subject. Patients are also encouraged to keep a diary, not only to keep a record of the day's events, but to put down the thoughts and feelings that have arisen as their power of judgement has come to life. Last but not least they are encouraged to take up art, and one looks at the pictures they have painted, listens to the music they have been practising if this is possible, etc. None of this should take as much time as private coaching would, and the important thing is that at least something is done in this direction.

Nor can physicians take the place of the clergy. Ideally one would work together with a priest who is able to go into the young person's longing for religious experience. One will find that there is sometimes a real hunger for ritual occasions. There will however be occasions when the therapist has to help out in the religious sphere. If a priest cannot be approached or at least not yet, suitable reading matter may be suggested and discussed with the patient.

In more serious cases where the longing can only be dimly sensed and cannot be put into words, there is little possibility of clear judgement and warmth of feeling. This applies particularly to young people whose developmental disorder has led to drug addiction. In cases like these, and with more severe developmental disorders, one cannot simply fill that inner emptiness with all that is good and beautiful and thus hope to strengthen the power of judgement and bring new life to their feelings. In fact one would achieve more or less the opposite. Unable to take in what they are offered, these young people fall into even greater despair and withdraw into opposition. It is necessary to use a very gentle approach, offering just a little food for the soul that is 'easily digestible' to arouse initial interest. Taking such a 'dietetic' approach, one avoids overloading the soul, for this may well react just like a stomach that is filled with too many good things.

It is a good idea to relate the things one offers to earlier interests in the patients' life. One may also encourage them to review the day's events at night, and this provides the basis for the diary entry. Looking back over their day, patients should ask themselves if there was anything that made at least some impression on them, where they were able to form a first opinion, and where they have been able to feel something. These are threads that can be picked up. Sentient life can be effectively stimulated by simple artistic exercises; patients should do these as rhythmically as possible, without any immediate need to achieve.

To bring the longing to birth it is first of all necessary to learn to ask questions. Many young people with developmental disorders (and indeed many adults) no longer know how to ask questions. The questions put by the therapist should encourage them to learn this skill again, but therapists must be careful not to have too much ready-made content, nor to have too many of the answers themselves. Answers given at the wrong time can stifle any question at birth and they will be rejected with even greater vehemence than the more neutral educational subjects. Questions have to arise before longing and interest can show themselves. This will also act against the set opinions and criticisms the intellect produces, a form of making judgements that presents a major obstacle in the promotion of development. One has to ask one's own questions before judgement becomes personal judgement and the road to perceptive knowledge opens up.

The kind of questions one might ask would be as to what they think is not right with the world, what changes they would like to see, what kind of work they envisage for themselves, what kind of partner they would like to have, and so on. If the line of questioning can be maintained for as long as possible one will find that it brings its own answers, or that the answer merely confirms something the young people have 'somehow' known already, something that may well have been the deepest source of their longing.

If neurotic traits of the kind discussed for the fourth seven-year period develop in the third seven-year period, and particularly in the second half of this period, the treatment outlined for the fourth seven-year period would be applicable, though with certain variations.

The above guidelines for therapy are complemented with medical treatment and remedial eurythmy. (Some details of this are given at the end of the next chapter.) Both methods act profoundly on the life of the body and its organ systems where the physical basis for the mental and emotional disturbances is to be found. This is done through the physical substance of medicinal agents or through movement. In cases of drug addiction, for instance, it is important to treat the liver which will often be found to be disordered. It will however be impossible to make real progress in inner development unless the ego of the therapist comes into play in the therapeutic approach that has been described above. Without such explicit psychological help there is little hope of achieving real results in the treatment of severe developmental disorders and of drug addiction.

Developmental Disorders of the Sentient Soul

If the developmental disorders of the third seven-year period continue into the period of the sentient soul they are given a special note by the ego that has now come to birth. The disorder may become more marked, and the ego will be more involved in coming to terms with it. This may improve the prognosis, and treatment can be more specific. On the other hand the birth of the ego may be made more difficult by developmental disorders in the sentient body and this could make the prognosis less good.

It is always the ego that determines the final outcome. If young people with developmental disorders can be addressed more as individuals, the range of possibilities will be greater. If the disorder is aggravated due to addiction or psychosis and the ego is more or less eliminated, the prospects of successful treatment are less good. The general rule is that the greater the duration of a developmental disorder and the later treatment is initiated, the worse is the prognosis. This should not inhibit therapists but rather spur them on, for they may well find that even quite serious disorders can surprisingly take a positive turn. It should however be remembered, in assessing the results of treatment, that young people with developmental disorders may also spontaneously achieve a late maturity.

Ego birth may also be weak and the sentient soul underdeveloped in young people whose puberty and development went normally in the third seven-year period. They will then still be caught up in the problems of puberty in their twenties and show the emotional instability and polarities of soul life that were physiological in the third seven-year period but are now becoming pathological. These young people have not 'taken the helm' and their course continues to be determined by emotional storms and the waves of sentient responses, for they are unable to steer their way through the *Sturm and Drang*. The healthy *Sturm and Drang* of a fully developed sentient soul bears the mark of egoity; here, however, the impression is one of uncontrolled urgency of desire. This may be called a persistently juvenile rather than infantile character. Neurotic subjects often show this and they will often look young for their age, sometimes for the whole of their lives.[32] ('Neurosis' is rather an unfortunate term; it is used here to avoid confusion and will be discussed later.)

This developmental disorder is partly due to the education young people receive at school and outside of school life during the third seven-year period; they are not sufficiently encouraged to form their own opinions and the will of the struggling ego has not been challenged the way it should be. Too few ideals have been implanted in the sentient body, ideals that are vital to young people and which might now give support to the sentient soul.

Compared to the third seven-year period, disorders of soul life arise much more frequently in the immediate encounter with the world during the fourth seven-year period. As Horney put it, 'the neurotic individual of our day and age' is produced.[33]

The Neurotic Individual and Anxiety

The problems of neurotic subjects often differ from those faced by healthy individuals in this day and age in so far as they are pathologically enhanced and transformed. Competition is one of the problems young people have to face when they go out into the world, and with it the fear of failure. Anxiety in the face of competition plays quite a role even in school life today, and many boys and girls suffer this particularly after the age of fourteen. It may get worse when it comes to immediate confrontation with the world.

Anxiety is a basic soul force. Essentially it is an emotion rising from the astral body and human beings have it in common with animals.[34] It plays an even greater role today than it did before.

Zeylmans van Emmichhoven wrote: 'It is no exaggeration to say that there are many aspects to social life that are governed by anxiety.'[35] With reference to neuroses, i.e. psychological states where relation to the environment is abnormal, Horney defined the term 'fundamental anxiety'. This provides the base, often established in childhood, on which neurosis may develop. This fundamental anxiety goes hand in hand with an 'all-pervading feeling of isolation and helplessness in a hostile world.'[36] In neurotic subjects it is greatly enhanced and may fill the deepest recesses of the soul. It can at any moment turn into fear, i.e. anxiety relating to a particular object, 'anxiety made concrete'.[37] The object of that fear need not be all that fearful, in fact; the essential element is the fundamental anxiety that will give rise to fear on the least occasion.

'Anxiety' and 'anxious' derive from the Latin verb *angere* which mens 'to strangle, constrict'. This relates to the fear that 'grips the throat', a sensation we are all familiar with. Animals take flight when this situation arises; the astral body takes the living physical body along with it. Humans are also capable of this animal-type reaction, or they may take flight, or rather withdraw, in another way. The ego, which lives in the blood, withdraws into the inner organism and the blood follows. This withdrawal may go as far as the heart. The face may grow pale with fear and so may the 'face' that the heart presents, for the coronary vessels at the periphery of the heart contract just as the facial blood vessels contract when we grow pale. Fear may cause human beings to 'grow pale' even in the heart, and this tends to make itself felt in a feeling of oppression.[38]

The connection between the heart and the ego—we have already discussed this—gives weight to the question as to whether the fear has also taken hold of the ego. If we consider the essential nature of the ego, which in the final instance radiates into soul and body from the world of the spirit, the answer is definitely no, certainly where the core of the ego is concerned. Only the astral body is able to have inclinations and disinclinations and also fear and anxiety. The ego may make itself subject to anxiety, and in that case the flight of the astral body becomes withdrawal of the ego. Its radiance is submerged in the life of the soul and further radiance may be prevented from reaching the soul life. The light of the ego is obscured by anxiety, its power is reduced; yet it is not filled with anxiety to the core of its being the way the life of the soul is. Under certain circumstances the life of the soul may be absolutely quaking with fear.

This view finds confirmation in the fact that the ego can do something with fear. Withdrawal of the ego may serve to concentrate its powers, and out of this the ego in the soul develops courage of a kind and strength that was not possible before the fear was felt. True courage always arises in overcoming fear, and this also involves the risk of foolhardiness developing as the opposite extreme, a danger that must be avoided. Between fear, where sensory perceptions made by the head, or mental images formed by it, constrict the life of the soul, and foolhardiness, which arises from the lower region of the organism and goes to people's heads, the heart's courage is born.

The more fear and anxiety touch the human heart, the more profoundly based is the courage we develop. This can be experienced in facing the fear of death, which is the truly heart-felt fear; if one is holding one's ground in the face of this fear the most profound courage for life may arise. Withdrawn into the heart, the ego has gone beyond death in the experience. 'Fear reveals our nothingness,' Heidegger said.[39] This ultimately holds true for the fear of death, for this is not fear of dying but fear of nothingness, fear of having the ego destroyed. This is where the impulse arises for the ego to remember its origins and discover that 'nothing' in physical terms means 'everything' in terms of the spirit. On the outside fundamental anxiety becomes fear, fear of life; inwardly it is enhanced to become fear of death and may thus mark the beginning of new life for the ego. The ego enters into the biography at its birth; it is given a special note in mid-life when individuals come face to face with their own death in the biography.

Behind that fundamental anxiety, however, original anxiety shows itself. This is an archetypal human experience as individuals go through the experience of being driven from Paradise, i.e. the original state of being bound up with the cosmos, and take the road that leads to the narrow confines of earthly existence. That original anxiety has become the fundamental anxiety of human beings who have now made their home on earth. Original anxiety therefore also points to the divine and spiritual origins of human beings whose egos can find a new relationship to that origin in overcoming fear and anxiety.

The Soul Life of Neurotic Subjects

Behind the two anxieties that arise from the fundamental anxiety—fear of death and fear of life–nothingness opens up. Outwardly it is experienced as the emptiness of the world, inwardly as the inner emptiness that arises when one has had the experience even of only partial destruction and the ego in the soul has been weakened.[40] The subject may also go to the other extreme and break out of that fear into foolhardiness or aggression. At this point the whole life of the neurotic soul, and not only sentient life, seems like an ocean that is much less able to come to rest than the sea of sentience. Once again the opposing principles active in soul life determine the issue. The sentient soul may either give itself up to the world in pleasure, or it may withdraw from it in displeasure. When the soul life has grown neurotic, excessive devotion to the world may alternate with excessive self-assertion. On occasion the opposing principles will be found to act concurrently. A fanatic lust for power may hide uncertainty and a feeling of inferiority, or vice versa.

It is possible to perceive not only the tendency to waver between one extreme and the other but also an effort of will on the part of the ego to establish equilibrium. The ego is also involved in setting up a facade with which to face the world. Activities like these do however mean subjection of the ego to elements of soul life such as fear of losing oneself or embarrassment in facing the world. In setting up a facade the ego is creating a deception; it is not acting out of the truth, though basically that is its aim. In wavering between two extremes the ego does not establish a true equilibrium but rather ministers to an excessive reaction in the soul, endowing it with personality traits ('overcompensation'). It has thus gone back to the state of personalized soul life belonging to the third seven-year period, a state in which it was not yet able to rise to self-attainment. The higher ego no longer attempts to speak through the earthly ego–the *persona*, as the masks worn by the actors of antiquity were called–but only anxiety, the soul's desire. The ego should have become the centre of the soul; instead it has come to serve as a mask for soul life. (This does not apply to a facade put up out of consideration for others.)

Repression, a process that relates to a reality in the life of the soul,[41] basically also does not arise from the ego or the kind of superego postulated by Freud.[42] The process of forgetting, which Steiner called 'a sequence of mental representations going to sleep', has already been discussed. This going to sleep need not always be spontaneous; it can be deliberately sought and brought about. The compulsive pressure produced by a complex full of conflict that has not yet been properly assimilated and wants to come up as a memory is opposed by compulsive pressure created by antipathy in the soul. This negative compulsive pressure results in the complex 'falling into a heavy sleep' again similar to the heavy sleep the whole human being falls into after taking a sleeping pill. The ego is involved in so far as it desperately tries to concentrate on other images in an effort to suppress the complex that would arouse fear or shame. This negative, suppressive pressure engendered in the soul need not only relate to a complex of images; it may also be directed against an emotion that is stirring in the soul and thus conjure up the opposite emotion. Thus a suppressed feeling of inferiority changes into a lust for power.

Repressions like these may be seen as the personal metamorphosis of a process that according to Steiner has involved the human race for generations. 'Human beings have repressed the drives, desires and passions they themselves have brought to life', pushing them

down into the 'subconscious that blazes up in the ether body.' ...
But the suppressed material also eclipses the divine principle that
lives in the subconscious.[40] And in the final instance that is also the
very source from which the longing for the divine wells up. Another
way of putting it is to say that the subconscious begins to expand
into a higher awareness, that a direct link is established to a super-
awareness where the ego can be consciously experienced.

People may try and kill their drives and desires by constantly suppressing them. Steiner expressly warned against this kind of false
asceticism.[43] He considered it important not to suppress powers
arising in the soul but to purify and transform them. Dull wrath may
be purified to become the noble wrath that arises in the face of injustice
or stupidity in the world. Transformation of such noble wrath will
give rise to love.[44] Love not merely wants things to be different in
the world or in one's own life, the way a noble wrath does; it can
be creative and bring about changes. Such transformations can only
be wrought by an ego that intervenes in the soul process. The power
to do so is gained in that the ego actively opens itself up to its spiritual
world.

If this does not happen and the ego remains subject to the ocean
waves of the soul, the ups and downs of those waves may turn into
a negative pressure that threatens to suck in the ego. The waves of
the ocean become a cyclone, a vicious circle in the soul, that no longer
has any apparent opening but narrows down progressively, drawing
the ego more and more deeply into itself. Horney has described such
cycles, and these are actually spirals, i.e. one emotion causes others
to follow in a reflex reaction and in the process returns to its starting point, but at a correspondingly lower level. A feeling of humiliation may give rise to the wish to humiliate others in order to get
rid of one's own feeling of humiliation. Then fear of retaliation causes
increased sensitivity towards further humiliation. This in turn
enhances the desire to humiliate others, and so on.[45]

The ego is however capable of intervening at any stage in this vicious
circle and breaking the sequence. Seeking self-awareness it may ask
itself: Why do I feel humiliated by this person? Did I perhaps
humiliate him? Am I inferior to him in one way or another, and is
that why I feel humiliated? Is there anything I can learn from him
that will help me to overcome the feeling of humiliation? Why am
I particularly sensitive in this area? Isn't this due to a fear of retaliation, because I myself have humiliated others? Do I feel a particular

urge (better than 'desire') to humiliate the other person because I
want to make sure he or she does not humiliate me?

The Inner Emptiness Felt by Neurotic Subjects
One aspect of the archetypal phenomenon of polarity in the soul is
that emptiness of soul arises, not only through a primary lameness
where activity of soul is concerned, but also through a primary super-
effort and hyperactivity; individuals who develop such hyperactivity
in soul life will run out of steam after a time and inner emptiness
will be felt. In neurotic subjects this is even more marked. Now the
emptiness of sentient life comes fully to experience; it may actually
show in the background even during the hyperactive phase and indeed
be dimly felt by the sufferers themselves. The process may finally
cause life to 'dry up' completely in the sentient soul.[46]

Conscious awareness of the inner emptiness may be greater in the
sentient soul than in the sentient body. The soul's longing for the
spirit, a spirit the ego wants to experience in the soul instead of the
emptiness, may also emerge more clearly. There is a feeling that the
very existence of the young person has its root, or ought to have its
root, in this experience. The Austrian psychiatrist V.E. Frankl coined
the term 'noogenic neurosis' for the syndrome that arises out of this
situation. He considered it a new kind of mental illness that was due
to 'spiritual stress' and arose in conjunction with extreme boredom
or in the final instance an existential vacuum. According to Frankl
this mental condition is very common today, especially among the
young. It was found–mainly in Europe–that 20 per cent of neuroses
were noogenic; on the other hand a statistical sample Frankl took
from among his own students in Vienna showed that 40 per cent
had experience of existential vacuum, and among those who heard
his lectures in America the figure was 80 per cent.[47] In this context
Frankl quoted a remark made by Boss, that boredom was the neurosis
of the future, adding that the future had already started.[48]

This 'vacuum neurosis', as one might also call it, first shows itself
in a negative mood of extreme boredom. Then depressive states
develop and finally a paralyzing lack of initiative, coupled with a
feeling of utter meaninglessnes. These conditions do not respond to
any of the conventional treatments.[48] It is evident that this fundamental neurosis of our time may also lead to drug addiction as
an attempt to fill that inner emptiness at least for a time. At the
physical level it frequently goes hand in hand with weak kidney
radiations.

Frankl considers the initial cause to be loss of instinct and tradition; these are no longer there to give meaning to life. He quotes Pfaender's statements that 'the human race has reached an evening twilight where values are concerned,' and that the generation that was growing up in this twilight had grown 'blind to values'.[48] According to Frankl there has to be 'return to the world of values', to 'an objective cultural and spiritual approach', which may be stimulated by logotherapy.[49] He bases himself on the longing 'deep down in our being that is so far from fulfilment that it cannot be for anything but God.' This emotional path of longing, of faith, was the only one that would lead to the reality of God; a thinking approach could not do so. Faith alone offered a free choice.[50]

Taking up the point of an evening twilight where values are concerned, one might ask oneself what the consequences of this would be in the souls of the young. To stay with the picture: Values are still visible in the twilight but they have become colourless shadows. It means that young people see them and take them in with the intellect, with the head, but that there is no colour to them. Blindness to values therefore does not relate to perception of values but to living experience of them. Longing and desire are there, but education has failed to transform them into genuine interest. Personal judgement activated by will impulses has not been sufficiently stimulated; there can thus be no inner response to the values that are perceived, and the soul remains empty.

Judgment is the faculty of thought that still needs to be acquired, for this alone enables the individual to make free choices and act out of freedom.[51] If people with a powerful and convincing personality address themselves primarily to sentience and feeling, to faith, this may indeed have powerful effects; in the long run, however, such an approach is not the right one for the problems that have arisen in soul development. A faith that permits certainty of feeling does not come at the beginning but at the end of the path that has to be followed today, a path where judgement can first of all provide initial insights and new sentient responses that serve as a secure base and thus convey awareness of the reality of the spiritual world at least in its first stage.

Therapeutic Aspects
The basis for therapy has already emerged in considering the pathology. If the disorders in the sentient soul have their root in

developmental disorders of the sentient body, the treatment discussed earlier will apply. If the powers of a healthy sentient body are not adequately transformed as they become part of the sentient soul, the ego of the sentient soul needs special consideration.

To begin with one should sit down with the sufferer and consider their biography up to the present time. With the 'proper application of psychoanalytical methods' that Steiner considered important, this form of history taking will uncover 'soul contents that have not been worked out' and which cause disease even at the organic level. According to Steiner, this would only lead to diagnosis, however, and not to therapy.[52] The process is nevertheless important as a prelude to the actual therapy. Steiner himself was speaking in this vein when he suggested that the destructive forgotten mental presentations deriving from earlier experiences should be 'brought to mind again, as this arouses the powers that lead to health.' He said that the patient's ego was too weak to bring those presentations to mind of its own accord and needed help to do so.[53'] Distinction must be made between these mental presentations and others that would and should be forgotten; these should be left in peace.

The first feeling of liberation which patients may have through this approach also comes when they talk to their therapist about current conflict situations they experience, and therapist and patient work together to discover the neurotic fullness or emptiness of a soul that is weak in ego. The support given to the weak ego by the therapist then finds an ally in the first beginnings of self awareness in the subject. An example of the latter has been given in the last chapter (vicious circle). This is also the time when the seven-year rhythm in biography and the crises arising in connection with this may be mentioned. It is important, however, that the subject really wants to take these steps towards self awareness; they must not be imposed. The ego of the therapist primarily influences the soul of the sufferer; relating to that soul in warmth of feeling, it stimulates its capacity for will and understanding, initially therefore addressing the subject's earthly egoity.

This initiates the actual therapeutic process; like any other therapy it is a process of self-healing, of healing brought about by one's own ego. Above the baser earthly ego is the higher ego through which human beings relate to the world of the spirit. Therapy in its proper sense should take this fact into account, as Steiner stressed in his lectures on psychoanalysis. 'It would be quite improper to treat anything that goes beyond the individual in an individual manner; those aspects require treatment in general human terms.'54 'Passing through hidden depths of soul we also reach the hidden depths, the spiritual background, of what is eternal and immortal in the external world.'27 As has been shown earlier, taking this as a starting point we can also meet the higher, immortal human ego that continues through repeated lives on earth.

The way in which we descend to the hidden depths of soul is important. The intellectual thinking at the daytime level of consciousness does not do justice to the life to be found at those depths; empathy is needed and 'real powers of imagination' that will produce a 'comprehensive picture'.27 (See also the chapter on Awareness Soul and Spiritual Training.) This, too is 'properly applied psychoanalytical method' and at the same time therapy. It is a route that will not only produce images of existing soul states, but release the ideal images held in subconscious soul life that have resulted from transformation of ideas 'from above'. The development of ideals that is an educational principle in preparing the sentient soul can also inspire therapy. Questions of the kind mentioned above, e.g. what one imagines the ideal occupation to be, already go in this direction.

The search for self knowledge, too, should not be entirely introspective, leading to broodiness. Humans have to 'realize that they are totally bound up with the rest of the world before they can say "Know thyself!"'46 It has been shown that knowledge of self and of the world come together in harmony as the awareness soul develops. The light of this soul illumines the development of earlier soul principles, and one may therefore start to bring those two aspects of knowledge in relation to each other during the seven-year periods belonging to those other soul principles.

These measures will be complemented by what has already been said on the subject of anger. Steiner called anger an 'educator of the sentient soul'. He said that anger calls up the ego, 'so that it may confront the outside world.' Impotence felt after anger encourages selflessness. The anger of the sentient soul should not be suppressed, therefore. In overcoming it we are able to transform it into noble wrath and into love. First, however, 'we must have something that can be overcome.'44 One should certainly consider providing the opportunity for anger in the course of therapy where anger does not arise spontaneously.

Treating Mind and Spirit

Catching up on education immediately takes us into the realm of common humanity, as does release of the longing for the spirit. When the ego is born in the seven-year period of the sentient soul, educational efforts at catching up may gradually be replaced by self-education. Individual efforts in this direction may indeed be given greater emphasis even during the time of preparation for this birth.

According to Steiner, it is a fundamental truth that anything missed in the years of one's youth is difficult to catch up on. But the possibility is there. 'It will then be necessary for the individual to enter very deliberately and in full awareness into a profound inner meditative approach ...' Steiner suggests entering into the experience of the great philosophies of the world as an exercise, for these take one to 'the great, all-embracing secrets of the world'; he also spoke of repeating the same prayers daily.[55] Meditative study of suitable passages is one approach to this, but the question also arises as to meditation exercises.

Steiner certainly suggested such exercises in conjunction with advice on medical treatment and eurythmy to a number of people suffering from physical and mental illnesses. Some of the exercises are such that anthroposophical physicians could also give them to other patients where indicated. Yet because the ego is struggling to attain to freedom such exercises should never be prescribed let alone demanded. They may be suggested, and many patients actually ask for such suggestions nowadays.

Meditation exercises initially are preparatory, or accessory, exercises for the three basic faculties of soul life performed by the ego—thinking, feeling and will activity. Thinking is prepared by the concentration exercise; in the same way the will exercise acts on the will, and the exercise for achieving composure on our faculty for feeling.[56] It will be for the patients themselves to decide whether they want to go on to the meditations proper after those preparatory exercises; it will also depend on the state of health in a patient's soul whether the suggestion to do so is made.

Health of mind and soul is a precondition for spiritual training in the anthroposophical sense. 'Unhealthy conditions in the life of mind, heart and soul will always put the seeker for higher knowledge on the wrong track.'[57] Meditations of this type may even be harmful for such patients, and so is reading in this field, which tends to be done in excess in these cases. Anyone suffering from mental illness should initially be strongly warned against either, even if they are sometimes asking for them in a way that almost shows a desire for dependence. At the same time we should not deprive them of the hope that once the conditions are met, i.e. once they have recovered, they may be able to follow such paths.

The preparatory exercises are much more likely to be indicated, and mentally ill subjects may find that these make them more self-possessed, active and composed. It is also important to encourage the soul to open up and to activate the ego to relate to the 'world', the 'spiritual world', whilst those preparatory exercises are being done. Goethean nature study will show the way out of chaos in the soul and let the first sentient responses arise in its emptiness. One thus directs the soul's attention to the shapes of clouds and of trees and asks patients to describe the impressions gained; this will encourage the processing of such impressions so that sentient responses can arise.[58] It is important that this does not stop at the early kind of undifferentiated emotive response that is primarily motivated by desire. 'That's simply beautiful!' Sensory perceptions can only live on in differentiated responses through which the young person relates to the world. Questions one might ask oneself are: 'Surely I feel something different when looking at a tree than when looking at a cloud? Both are beautiful. But why is it that I feel different with the one compared to the other?' Repeated practice of careful observation will lay the foundations for differentiated responses.

When one is thus endeavouring to bring soul qualities into an impoverished life of the senses, it is also important to suggest reading matter that will stimulate the neurotic individual's power of judgement. We have seen that it needs thought-out judgement to transform dim desire into the bright interest that human beings need in the life of the senses. Thought-out judgement is strengthened by philosophy and particularly the theory of knowledge. Rudolf Steiner's works relating to the latter are specially recommended for this purpose, particularly his *Philosophy of Freedom*.[59] Thinking about thinking is learned in the process, and this helps to develop the kind of clear judgement that may well be experienced as a first revelation of the spirit. It is vital that one feels something in reading such works, that one enters into the experience, i.e. that judgement comes alive and the thoughts one has taken in are not merely held in the head. If the latter does happen, care must be taken to ration the study of works on the theory of knowledge during the fourth seven-year period. Steiner's warning that 'constant effort to develop one's thinking' can cause the life of the soul to 'dry up' specifically applies in this case.[60]

Art has the most immediate influence on the sentient soul. When art is taken in, it is made easier for the weakened ego to assimilate those impressions because they are themselves the outcome of assimilation. Exercises in one of the arts involve following the therapist and following the laws of the material and of form. Imitation and authority, two principles that have not been sufficiently brought to bear earlier on in life, come into effect in a right and proper way in the sphere of art.

This is also the sphere of the instinct for play, and through this freedom can be experienced as a balance is achieved between opposite poles. Combining material and form tendencies in harmonious ways, gaining self expression through the material and entering into the laws of form in a living way, patients gain strength in their own middle sphere that is struggling to gain freedom but has been weakened by chaotic excess or by emptiness in neurotic subjects. The beauty found in art lets the spirit come alight in sentient life. One's own self is not merely given up, as is necessary in the search for truth, but 'aesthetic judgement gives our own self back to us as its gift.'[60] This is particularly important for young people whose egos have been weakened; impressions gained in art also grant them the experience of their own egoity.

Medical treatment addressing the sentient body and the sentient soul has its basis in treating the kidneys. (The connection between the kidneys and sentient life has already been discussed.) Copper, the kidney metal, supports a healthy warmth of sentient life; it is indicated if the astral body is overactive and there is a tendency to excitement, and when it is lethargic and there is the feeling of emptiness. Iron on the other hand helps young people to complete their incarnation into the physical body and into life on earth and to overcome their fear in this respect.

Remedial eurythmy would focus on the consonant B and the copper vowel A (as in father). B has enveloping qualities, and a young person who has been cast upon the world and feels naked would benefit from this. A is the vowel for awe; the soul learns to open up to new sentient responses. (For further details, see the chapter on Art Therapy.) The *Kibitz* (lapwing) M, with the up-starting leg caught and held in a harmonizing M, helps to counterbalance the uncontrolled impulses rising from the lower sphere of the organism, particularly the sexual sphere. Here a special silver preparation is the indicated medicine. Neurotic instability in the sphere of feelings, vacillating between sympathy and antipathy, needs to be addressed via the lungs, using mercury. The consonant M and the mercury vowel I (as in deep) also help in this case, with the latter challenging the ego as it seeks to find the balance. The iron vowel E (like the first part of the ai in pain) acts in the same direction as iron itself. (One would not, of course, use all these exercises and medicines simultaneously.)

In principle, it should be noted that none of these medical and eurythmy treatments can take the place of individual effort. They merely open up the way for further development; the essential principles of human beings are challenged and strengthened, but they must walk the path themselves.

Developmental Disorders of the Intellectual or Mind Soul

In this case, too, one would first of all consider any existing developmental disorders relating to the earlier seven-year periods. If still active, these will be again be more in evidence and act as obstacles to intellectual or mind soul development. On the other hand the ego is coming in more strongly at this point and may arouse an new impulse in the individual concerned, an impulse to do something about the underdevelopment of soul life.

When the sentient soul has been fully developed, the problem of the birth and development of the intellectual or mind soul arises in around the twenty-eighth year. The pathological phenomena that show themselves in soul life at this stage always make it clear that, in spite of initial attempts made in the way described above, people find it difficult nowadays to take the inner path in mind and heart and develop their own thoughts in the search for truth. They feel unable to go beyond the surface both where their own existence is concerned and with reference to the world. The tendency is to blame the rush and hurry of modern life which is said to prevent people from finding themselves. At the same time they can no longer really relate to the world, and they find that there is not enough time to think. Yet compared to earlier on in life, this becomes much more of a need now, a need people feel cannot be met for external reasons. Poverty in the life of mind and heart goes hand in hand with this. The people concerned seem cold, and one experiences an inner emptiness in them that has to do not with sentient responsiveness but with mind and heart. It means that individuals still show initial, rapid, and sometimes passionate, sentient responses to a human encounter or something experienced in nature or in art; but all this remains very much at the surface and does not reach the inner life, where these things should be pondered in mind and heart. As a result those experiences tend to be short-lived, leaving few traces in the soul. Martin Luther King called it one of the great problems of the human race that we suffer from poverty of mind and heart.[61] This is one aspect of the problem situation one frequently meets in people in their late twenties and older.

The other aspect, underdevelopment in independent thought, is also widely apparent in adult life. Rather than forming one's own thoughts in the process of due reflection, there is a tendency to reproduce things one has heard or read somewhere. The extreme case of this is 'thinking in slogans or headlines', often quoting word for word from certain papers. When faith forms a union with thought contents that have been adopted and not properly assimilated, dogma results. Not only religious contents but particular philosophies of life may become dogma, and people will become the slaves of such things–sometimes without being aware of it–or present them as their own ideas.

During the epoch of the sentient soul not everything has to be fully assimilated in thought; the soul shows a living sentient responsiveness that still enables it to adopt the thoughts of others. If this continues into a later epoch, however, such adopted thoughts–and the original life of the sentient soul–die and become slogans or dogmas. Things go even further if thinking is in platitudes, feeling becomes rigid convention and routine governs the life of the will. Since the end of the 19th century, this caused a 'spiritual ice age to develop in social life.'[62]

The adult world carries the responsibility for this, and young people have to suffer under it until they reach their late twenties. They rightly rebel against this but finally, having failed to weather the crisis, fall prone to it themselves.

The type of character that develops is someone who externally lives a decent and orderly life, following his chosen path in life more sedately than heretofore. It appears that one has grown sensible; one has come to accept. Things are no longer felt as passionately as before, and should desire arise the entertainment industry will provide, whilst television and magazines meet the longing to learn more. One is more or less turning into a mild form of bourgeois conformist.

Others, who have special gifts, continue in life by consolidating and perfecting everything they have been able to acquire because of those gifts. Outwardly they are successful; life goes on and they become famous and respected. But does this also imply further development? Is anything essentially new coming into their creative work and into their lives?

No Further Development after the Age of Twenty-Eight?

Some individuals come to a definite stop in soul development when they reach their late twenties. 'People who have previously shown special gifts, with a creative facility that seemed inspired, will suddenly grow shallow. Some who have been brilliant writers in their youth often lose there creative potential at this point...People for whom a great future was predicted come to nothing. Promising talents lose all originality.'[63] Sheehy comments that now 'a valiant if awkward struggle for ourselves and against our inheritance' is beginning. She says that efforts are made to establish a new bond between inherited and individual characteristics.[64] That is why in mid-life Goethe's Faust said that anything inherited from one's forefathers had to be actively made one's own.

With reference to those thoughts spoken by Faust when he was about forty years of age it should be noted that this experience of inner development having ceased at around twenty-eight years of age, which is a crisis point in many lives, may be gone through unawares by others and only come to awareness in mid-life. It may however make itself felt in so far as pathological symptoms develop at around twenty-eight years of age, symptoms we shall discuss later. In every case where development comes to a stop the defects described above will develop, though in some cases they do not show themselves clearly and do not take effect until later in life. In a gifted individual whose influence extends far into the periphery, cessation of inner development and its sequel may have tragic consequences for the world.

A brief biography of President Woodrow Wilson may serve as an example. Steiner made special reference to this.[65] In his young days, Wilson developed quite a few ideals that may be classed as significant for humanity. He was a persuasive speaker, and the youthful elan of his sentient soul would sweep his audiences off their feet. Wilson was appointed Professor of Rhetoric, he became President of an American university and finally President of the United States. In adult life a liking for discipline and organization combined with the ideals of his youth, but there was no further development of those ideals and they were not related to the social structures pertaining at the time. This thinking was at an outer level and did not lead to new insights. Mind and heart never came to flower, and inner relationship with other human beings, and with reality, failed to develop; yet for the time being he was still a gifted speaker.

Thus inner contact was poor, and this went increasingly hand in hand with intolerance. Wilson failed in his work at the university and became President of the United States. In that role he preached an ideal of peace that had grown abstract, and his sympathies for the Allies resulted in the USA getting mixed up in the Great War. When that war came to an end he proclaimed 'peace without victory', yet the Treaty of Versailles had the opposite result. The world also turned many of his 'fourteen points' inside out, self-determination, for instance, no further annexation nor secret diplomacy, and disarmament agreements. Wilson died at the age of sixty-eight, a disappointed, prematurely aged man haunted by delusion.[66]

When people in their early thirties are onesidedly living at a superficial level where their inner life is concerned, this indicates that the intellectual or mind soul has only weakly come to birth and that development has been minimal. Soul development remains more or less at the level of the sentient soul, yet on the outside life continues. The powers of the intellectual or mind soul show themselves in a desire for order, for consolidation, organization and perfection–as in the case of Woodrow Wilson–, but those powers are exhausted in turning to the outside world and do not lead to inwardness. Encounters with the world were rightly experienced at the soul periphery during the life epoch of the sentient soul, with the peripheral nature of many encounters offset by their immediacy. Now, however, a soul life that remains at the periphery gives quite a different impression: The sentient soul attitude is maintained but no longer has its original vitality. The ideals of such a soul become empty, routine concepts, and a thought content that was once taken in and brought to life turns into rigid slogans and dogma.

To see it as a picture: The mountain stream loses its fresh virgin nature. It either threatens to lose itself in a morass, or it is tamed and has to serve industry, carrying away its effluents. In the outside world we have environmental pollution; in human souls 'pollution of the ego' occurs, and the self can no longer hold its own as the world takes its course and all is hustle and bustle.[67]

We can thus only speak of development coming to a standstill in the late twenties in so far as inner development comes to a halt for the time being at the level of the sentient soul. Outer development progresses, however, and individuals will try and maintain the status quo for the soul life they have developed. Yet how does the soul react to this?

Premature Ageing and the Desire to Stay Young

Once again the soul finds itself subject to head impulses. Anything that has arisen and taken shape in the process of development so far is continued with; the intellect of successful individuals makes sure that their outward appearance is as perfect as possible, 'selling' to best advantage. The reflective approach that led up to the birth of the intellectual or mind soul now serves to organize things for the future, but this is merely in continuance of what has been in the past and there is no new development. The plateau that has been reached does not turn into fertile soil where new things may grow.

The particular soul aspect that essentially relates to the future in its desires and in a life of the will that is now assuming more definite contours, turns against a tendency like this is inimical to the future. A disastrous dichotomy develops in the personal biography. The intellect projects the past into the future (life has to go on, after all, and one has to learn to cope); yet in its will and feelings the soul directs its longing away from the future and towards the past. People try to escape the premature ageing process in their souls–and sometimes also at the physical level–and attempt to recapture parts of their youth that have been lost and others that are increasingly slipping from their grasp, wanting to put new life into these.

This may take various forms. Efforts to look as young as possible and behave in a youthful way are common; from there the road goes via hormone therapy to the different forms of addiction described earlier on, addictions based on sexual excitement, television and extreme devotion to sport. Drug addiction and alcoholism in particular are given new impetus. There is a desire to feel young again, if even for a moment, and it is met by 'buying momentary happiness'. It cannot be denied that intoxication can sometimes prove stimulating and that the sobering-up process that follows may cause one to reflect. Yet ego activity is weakened each time this happens, development goes into reverse, and no real rejuvenation can be achieved by taking this route.

The desire to be young is so powerful that it shapes the whole of our present life-style. Modern society makes a fetish of youth.[68] The general slogan is 'You've got to be young to be with it.' Bodamer notes an absolute 'compulsion to be young' from which none dare opt out.[69] According to Guardini modern life is 'essentially governed by people who have not grown up.'[70]

Some like to pretend to a youthfulness they no longer have, particularly face to face with the world. Looking at paintings or photographs of leading figures from earlier times, one will again and again note a serious composure that speaks of a sense of responsibility and would have inspired confidence. Today, however, 'keep smiling' is the key word on almost every public appearance or presentation in the media. The general opinion is that you cannot entrust your future to someone who does not present a smiling and preferably youthful face.

How do young people see this trend? Do they want to see older people behave just like themselves? Not only sons and daughters feel some kind of pain when their parents go to extremes in putting on youthful airs. Do they feel that they are being caricatured? Whichever way this may be, they desire to meet individuals who have truly grown old and who have endeavoured to gain wisdom in old age. From them the young would also be inclined to accept occasional advice. Instead they find that they are being criticized for doing exactly the things their elders would rather like to do themselves but find they are not longer really capable of doing.

The battle of the generations is no longer so much between young and old, but rather between young people and those who do not want to grow older and desperately hold on to their youth.

This chapter has shown that in the long run development can never stand still. It either moves ahead towards its goal, or it becomes retrogressive. Another possibility is that it splits in two directions: On the one hand development continues in all outer respects, but on the other hand pathological regression occurs in the life of the soul. The impression is that regressions of this kind are on the increase and are more and more frequently resulting in illness. Even settling down and becoming bourgeois is getting less and less likely to ensure healthy development where the outer aspects of life are concerned.

The General Human Background
The developmental disorder we have just been discussing is so widespread today that every one of us should first of all ask himself how things are in their own case. It will be found that today everybody is more or less struggling against stagnation or regression in soul development, a problem that started in the late twenties and has continued through the years that followed. Surely we all often feel tempted to stay at the surface, not to make the effort to think deeply but rather to repeat something we have heard or read, though we have not pondered it in our hearts and inwardly assimilated it. Surely we have to discover over and over again that inwardness, a warmth of feeling for the other person that arises from within, often calls for effort and then again further effort and has to be deliberately cultivated.

According to Steiner, 'the sentient soul is most strongly developed' in people today.[44] The threat of soul development coming to a standstill affects primarily the intellectual or mind soul. Underdevelopment of the sentient soul and sentient body can however also be observed. It is far less common than underdevelopment of the intellectual or mind soul, and in many cases the deficit is to some extent made up later in life, something that cannot be said for these disorders when they affect the intellectual or mind soul. The latter are a problem on the general human scale even in so far as they are so common, and that no doubt is also the reason why they are not given the attention that is due to them in conventional biographical studies.

According to Steiner, the threat of soul development coming to a halt points to a connection with human evolution as a whole.[65] Going back to earlier times one finds that development did not come to a halt then but continued right into old age. People were capable of further development and educable for the whole of their lives. Unlike now, old people were revered because they had grown wise in the natural course of development under the guidance of the gods.

Until the middle of life, the vitality of life that was in the ascendant in the body, which had a connection with the cosmos, had borne the soul within it. After that, the ageing body released the soul: Withering of the body led to a flowering of the spirit, and this took the soul along with it. The gods raised the spirit to the divine world, and when death came that world was wholly open to the spirit.

Humankind has been getting younger and younger in this respect as its evolution progressed. The time when soul development came to a halt was coming earlier and earlier, unless the ego had come awake in the meantime and assumed the task of taking soul evolution further. The time when healthy dependence of the soul on the living body and on the gods ended was coming earlier and earlier. This meant that the time when the divine world would consider the ego worthy to take charge of soul development was coming earlier and earlier.

During the Greek epoch of civilization, soul development was still spontaneous until the middle of life was reached. Up to that point people felt entirely at home in the natural world and the elements of the cosmos; life was in the ascendant in the body and related human beings to the natural world and the cosmos, which to them were peopled with many different gods. Death normally comes to awareness once the middle of life has been reached, and the previous, Egyptian civilization had a matter-of-fact relationship to it. In ancient Greece, only those who still followed the mysteries had an understanding of death. For ordinary mortals the phrase 'Rather a beggar on earth than a king in the world of shadows' was coined. To the Greeks, therefore, mid-life was the high point, the acme, of life, and fortunate indeed were those who died at this point.[71]

After this, the end point of spontaneous soul development continued to come earlier and earlier and this was now in the first half of life. Now the living body was no longer releasing the soul the way it had done before; instead its life, still in the ascendant, threatened to 'overpower' the soul, so that human beings could no longer look up into a divine world merely on the basis of spontaneous development. Materialism and atheism were able to take root in human lives. When this situation first arose for humankind–in the 33rd year of its development–Christ came to earth and in a life span of thirty-three years established the preconditions that would allow human beings to relate to the divine world in a new way. Speaking of the moment when his researches in spiritual science led him to this discovery, Rudolf Steiner confessed that he had known few moments when he had felt so profoundly touched as at that particular moment.[65]

From the time of Christ onwards, the power of Christ, who had gone through death, was an element in human evolution and united with the human beings who were open to it, irrespective of religious confesson or adherence. It stayed with humankind until the beginning of the present age in the fifteenth century when humankind of its own accord reached the soul age of twenty-eight. This marked the end of the Greco-Roman period, during which the intellectual or mind soul for the whole of humankind had been fully evolved, as we can experience today if we enter into the intellectual thought of Greek philosophy and the qualities of mind and heart of the Middle Ages. Then mental horizons began to expand on earth through inventions and voyages of discovery and the epoch of the awareness soul began. This may be seen as a counterbalance to the progressive juvenility of the human race and the increasing problems and demands due to this. According to Steiner, spontaneous soul development now comes to a halt in the twenty-seventh year. One epoch of civilization may be considered the equivalent of a seven-year period in the life of a human individual; it takes about 2160 years, so that 300 years correspond to one year in a seven-year period. In 1981 [the year when the book was first published in German, translator], 568 years have passed since the new age began in 1413. This means we are reaching the end of the second year in humanity's seven-year period of the awareness soul, which coincides with the seven-year period of the sentient soul (21-28) in the life of the individual. Development thus comes to a halt in the twenty-seventh year, or if we are very accurate, in the twenty-sixth year of life.

The time given earlier, 'in about the twenty-eighth year' may thus be more specifically be said to be in the twenty-seventh year. It means that problems arise even where the complete development of the sentient soul is concerned. This is evident in many case histories, where the twenty-seventh year stands out by itself, but has also become the general situation. Most people still have reasonably well developed sentient souls, but some young people stop at the sentient body in their development. This anticipates future times when much more serious mental and emotional disorders will develop unless fundamental steps are taken to prevent this. In the above-mentioned lecture, Steiner spoke of an 'epidemic dementia praecox' (an earlier term for schizophrenia) that may arise once development threatens to come a halt in the third seven-year period. Under given circumstances people might then only reach 'the maturity for life of seventeen-, sixteen- and fifteen-year-olds'. (See also the chapter on schizophrenia.)

This negative aspect is counterbalanced by the positive fact that the development of humanity's awareness soul progresses and individuals are given increasingly greater opportunity to let their egos take their development forward.[72] The light of the awareness soul illumines the whole of soul development today. If it shines not only in the world but warms and illumines the ego in the way described above, we may see it as a ray of the light of Christ that helps and heals. If however the opportunities now opening up for the ego are not taken up, if the longing for the spirit that is growing all the time is not released as awareness develops, then even the souls of the young will receive less and less inspiration from the physical body in the course of development and will be 'overpowered' instead. Pathological dependence on the physical body will be the consequence, something we have already seen in the retrograde development connected with addiction.

Physical Aspects
Earlier on in this book, the fact that the intellectual or mind soul emerged out of the ether body in the second seven-year period pointed to the role played by the seven-year periods of physical development. The developmental disorders of the intellectual or mind soul also have one of their roots in the second seven-year period. If development of the ether body during that period is disrupted, this leads not only to the direct consequences for the development of the sentient body that were discussed earlier. Repercussions will also be felt much later, in the fifth seven-year period, when there are problems in developing the ability to think for oneself and to become more inward in heart and soul, for these have their basis in the ether body.

To find the road to the world out of the inwardness of a thoughtful heart requires more courage than it does when this is done on the basis of sentient life, for in the latter case the world still encourages and supports our efforts to reach out to it. The soul also needs more power of will and more initiative than it did before. It is due to this that one sees 'cowardice' and 'indecision' in people who have reached the end of their twenties and in whom the development of the ether body had been impeded during the second seven-year period.[55]

Timidity and indecision may later turn into existential anxiety and paralysis of will. (This will be discussed in more detail at a later stage.) In many of these cases the history will show incidents in the second seven-year period that did not appear to have a major effect on soul life at the time but in fact went deep and interfered with the ether body's development. (Physically this may show itself in a general anabolic weakness.) The ninth year of life seems to be particularly important in this respect. Problems arising with a move to a new house or school, death of a parent, or serious illness suffered by the child at about the age of eight may have the above-mentioned consequences for the soul life of the fifth seven-year period, unless measures are taken to prevent this.

The timidity and disorders of will life of adults may lead to depression, but more commonly lay the foundations for the fundamental anxiety and weak-willedness that are characteristic of neurotics. We have seen the weight that attaches to anxiety in modern people who are in their twenties or older, and it is evident that similar weight attaches to one of the roots of this anxiety which lies in the second seven-year period. There need not have been any particular incident during that period; the sins committed in upbringing and the education received at that time have had their effect on the ether body which had come free at that time and have led to the above-mentioned problems in the fifth seven-year period. This relates above all to undue depletion of the creative powers of the ether body at school An additional factor may be traumatic events of the kind described above; these may loosen the creative etheric powers. In either case, the ether body loses its capacity for being formed or moulded, and this partly inhibits the metamorphosis of its powers into the powers of the intellectual or mind soul.

As already mentioned, physically the fifth seven-year period is characterized by perfecting and consolidating what has been achieved before. At this stage the life of the soul should separate from the ascending curve of physical development and begin to develop independently. Soul development is not taking its normal course if it does not take charge of its own development at this stage but continues to depend on the physical body. At this point–i.e. before the midpoint of life has been reached–a process sets in that in the words of Steiner may be epitomized as follows: Soul development is no longer 'inspired' but 'overpowered' by the body. The physical body imposes its own developmental processes on soul development.

To begin with the pathological effect of this does not show itself clearly in individual souls. It may be perfectly justified in consolidating what has been achieved so far–for a time. The whole tragedy of the process will only show itself in midlife, when the soul does not merely stop at consolidation but begins to follow the descending curve of physical development.

Continued bonding of the soul to the physical body in around the twenty-eighth year also has consequences for this body. If the powers of the intellectual or mind soul cannot come to full development in the world they stagnate and this will cause disruption and disease in the physical body. Again one sees the retrograde development in the physical direction that has been discussed above, and it is easy to see that it will primarily affect the organs in which development had originated. The system comprising liver and gall-bladder has been shown to provide the basis on which the intellectual or mind soul developed, and one can therefore see why depressive states develop. In this chapter, we are mainly considering neurotic syndromes, and there is a characterisic neurosis relating to the intellectual or mind soul that is connected with the heart, which is the second organ to have significance in this context.

Neurocirculatory Asthenia
This condition, also known as cardiac neurosis, effort syndrome, or functional cardiovascular disease, combines all functional heart symptoms that have their origin in soul life. It is present in thirty to forty percent of all patients with heart disease. Patients presenting with this condition will first of all complain of physical heart symptoms. Briefly, they will report nervous palpitations with accelerated and sometimes irregular heart action. They will also complain of cardiac pain that may be darting, burning or of some other kind, and a feeling of oppression in the heart region and in the chest that may also involve breathlessness. The feeling of oppression, and in fact the palpitations, may go hand in hand with the cardiac anxiety we discussed earlier, an anxiety that may go to the extreme of becoming a fear of death. These patients are haunted by fear of death due to heart failure, although there is no organic basis for this fear (cardiac phobia).

Having excluded organic causes, one must go into the psychological history. It will be found that there has been a time when the feeling heart suffered, and this may have been on a single occasion or for an extended period. Partings appear to play a particular role: a 'heart-rending' separation that may have taken the form of the death of someone close to one, or indeed being present when someone died of a heart attack–any of these may have been the initial event. These are experiences that 'go to the heart' and always hold an existential threat, and any kind of experience may affect the heart if this particular nuance is present. [73] Concurrent physical stress due to overexertion will above all affect the coronary arteries and cause the above-mentioned pain and oppression, with the pain sometimes radiating down the left arm.[74]

The personal history of soul development will provide first-line answers to the question as to why some individuals develop neurocirculatory asthenia when they have undergone the kind of experience described above, and others do not. Some of these patients show a dependence on others that may in extreme cases be a downright 'symbiosis'. Separation from a domineering mother or from a partner will trigger neurocirculatory asthenia in individuals who are unsure of themselves and easily excited. This generally comes to expression in cardiac phobia with paroxysmal tachycardia, fear of death and fear of other attacks.[75]

Patients with neurocirculatory asthenia who suffer above all from oppression in the chest and even anginal attacks for which no organic cause can be found (vasomotor angina)[74] may show the kind of personality structure frequently seen with coronary artery disease. Unlike the type of patient described above, these patients have usually been very strictly brought up; they are excessively controlled, tend to rationalize and constantly escape into 'activity and achievement'.[76] This used to be called the 'manager type', caught up in the rush and bustle of life, but it is now realized that the situation applies to others as well. In patients who are in the habit of ignoring or minimising their symptoms, spasms of the coronary arteries may be an early symptom of organic heart disease which in the final instance may lead to myocardial infarction.

These two syndromes–mixed forms are also known–point to an upset balance between diastole and systole, expansion and contraction. In the first syndrome, where cardiac activity and emotionality are enhanced, the lower pole as the source of diastole is too strong. In the second syndrome, where there is a tendency to coronary spasm and even anginal attacks, the upper pole that governs systole predominates. Patients in the second group tend to rationalize and cling to their work, whereas those in the first group are more liable to give themselves up to their emotions and gain release from them through aggression. Both are subject to the fear of death, fear of losing their egoity. And both experience something that is indeed real, for the ego is indeed meeting with obstacles in its efforts to establish equilibrium in the heart. The functional disorders that arise through this can also cast the fear of annihilation on the soul.

Any kind of human experience that interferes with ego development out of the heart also will not let the intellectual or mind soul grow really heart-free. This starts when individuals are unable to separate from their mothers or from other people. Symbiosis then becomes an obstacle to the ego of the heart and this remains caught up in the organ in a way appropriate only to childhood. Subsequent separations cause the ego's powers of love to be held back and stagnate in the heart because they cannot find fulfilment in the world. Flight into work and constant busyness are no solution, for they will not resolve whatever weighs on people's hearts so that it becomes healthy activity.

These and other existential problems weigh on people's hearts even in their twenties. According to the statistics, the incidence of disease is highest between the late twenties and midlife or the fortieth year. This is mainly the time when the intellectual or mind soul should develop. Initially, therefore, the heart is driven to restlessness and into spasm by the powers of this soul, because they are prevented from gaining their freedom. (Concerning awareness soul and heart, see the chapter on this.) In my experience, the realization that the heart is suffering not merely due to external factors but that this is a struggle in which one's own soul is involved, a crisis in personal development, has helped many patients with neurocirculatory asthenia to cope better with their symptoms.

The crisis is not entirely personal, however. This type of disease is constantly on the increase–it is currently estimated that $2-5\%$ of the total population suffer from it[73]–and this points to the crisis faced by humanity. Being the ego organ, the heart perceives the threat of stagnation in soul development; in the light of developmental problems, this is what lies behind the patients' fear of a heart attack. Individuals who have recovered from neurocirculatory asthenia may feel an impulse to devote themselves to their tasks with greater courage and love than before and continue with their previously inhibited soul development. The illness may then be seen to have had meaning.

Therapeutic Aspects

A review of the therapeutic guidelines and experiences given so far will show that the work done in the third and fourth seven-year periods always took account of the next seven-year period. Developing the interest and power of judgement of individuals and implanting ideals in their souls in the third seven-year period one is also addressing the ego of the sentient soul. The training of thought and assimilation of inner experiences in the fourth seven-year period also prepares the way for the intellectual or mind soul.

If sentient soul development is on the whole complete, greater emphasis may be put on training thought and meditative inwardness, and patients may be expected to show greater independence. If they are willing and able to do meditative exercises they should be encouraged in this. Space is now being created in their mind souls and meditation will therefore be more successful. The principal line of approach arises out of the primal educational element in the intellectual and mind soul: Truth is the educator. In the search for truth-and not in the mistakem belief that one possesses it-the ego gradually wants to grow beyond the sphere of sympathy and antipathy. This also provides a special opportunity to train a systematic self knowledge that is now coming to be more aware. Many of us will find it most instructive to ask ourselves, as we form a particular opinion or come to realize something or other: 'How much of this reflects my own personal sympathies and antipathies?'[77] This applies particularly to the way we judge people, for in that case our positive or negative judgement often pretends to a high degree of objectivity or spirituality that in fact does not exist. According to Steiner, our own actions may also appear to be motivated by 'esoteric purpose' when in reality they merely serve to satisfy egotistical impulses. Such 'self justification' may sometimes cover up elements that 'rampage and rule in the depths of animalic life.' It may happen that such an individual has dreams of being pursued by wild animals, and moral interpretation of such dreams can lead to further self knowledge. Steiner suggested that dreams might be usefully interpreted by taking this approach.[78]

It will of course be important to avoid any kind of 'transference'. In the final instance transference can only be meaningful if the ego takes the lead, and when it is a matter of gaining self knowledge only the individual's own ego can do so. Patients who are at all capable of this should therefore be allowed to follow their own path to self knowledge. Again patients should feel that the physician is not sitting there as a 'demigod in white' but is also following the path and will be their companion for part of the way. It means that physicians, too, need to follow the path to self knowledge in the sense we are speaking of here, and as far as possible they should also have done the exercises their patients are doing.

More than ever it is important that the search for knowledge is not limited to the patient's inner life. Growing and expanding insight as well as interests and opinions should come to embrace as much of the world as possible. This will on the one hand be achieved through suitable reading matter and on the other by thinking through the events of the day. Even during the fourth seven-year period, the sentient soul's review of the day's events may be followed by something we may call a 'reflective review'. Once the day and its events have passed–and not before–we may have our own thoughts about them at night. We may ask ourselves what might be the possible significance of some event for our lives, or whether one was doing the right thing in a particular human encounter, and if not, why not. Why does one feel sympathetic towards a particular person? Could this develop into a real relationship, or do we merely want to see ourselves in the other person? Why do we feel antipathy towards another person? Does it really serve as a warning that we should not make contact with that person? Usually these are exactly the people from whom we can learn something, and yet we shy away in antipathy. In this way sympathy and antipathy can become organs of perception that lead us to other people. Having come to this kind of self realization we should then go one step further and ask ourselves: 'Looking at life so far, what do I owe to others and who are these others?'[79] This immediately makes us expand into the social sphere.

Nor should we allow patients with neurocirculatory asthenia to remain limited to the personal sphere. The first step will usually be to explore their life story with them to discover what weighs on their hearts, what has once been dear to their hearts and has now gone into spasm in their hearts. However, having enquired into their personal goals, one will also ask what they used to read in the past, and if there are any intellectual and cultural interests that they have given up under pressure of time or due to pressures created by their illness. We may advise them to take up some of those interests again and it may be found that the thoughts they have after reading some worthwhile book are more mature than they were before. Therapist and patient together move actively in the life of the patient's emerging intellectual and mind soul.

Nevertheless it will be found–particularly in people suffering from neurocirculatory asthenia but also in other serious cases–that the patients' egos and intellectual and mind souls are not getting active and cannot bring the therapist's suggestions or the patients' own resolutions to realization. The therapist will get a feeling that those souls and egos are held in spasm in the physical body and heart. Here and in other cases medical treatment is essential. It shortens the period of psychiatric treatment required and in fact will be needed to make it bear fruit on occasion.

The principal medicament is gold, the metal that combines maximum contraction, as is evident in its weight and density, with a maximum expansion that comes to expression in luminosity and the subtle 'radiation' of its physical substance through the whole world, so that a harmonious balance exists betwen contraction and expansion. Gold has a direct effect on the disturbed interplay of systole and diastole in neurocirculatory asthenia. It also takes effect wherever the ego is caught up too deeply in body and world, or where it has already partly withdrawn from the living body and from life in resignation. Gold helps the ego to reestablish a harmonious relationship with body, soul and world.

In remedial eurythmy, the *Liebe-E* [the E of love–with the E pronounced like the first part of the ai in pain] exercise is gold activity expressed through gesture. As the arms open up and reach out to the whole world, diastolic experience is touched on; it is brought back to the inwardness of systole as the hands are crossed over the breast. If the exercise is repeated a number of times this evokes a mood of balance that is of definite therapeutic value, with a feeling for the 'breathing' of love, where inside and outside, above and below come together in harmony. Again and again the extended arms receive, and what has been received is taken into the heart in the gesture of contraction.

The O gesture acts via the liver and helps to round out a mind and heart that was revolving upon itself; it combines well with the metal tin, for this specifically helps to maintains balance in the liver system.

Rejuvenation through the Spirit
The approach to treatment outlined above–it does of course only represent a fragment–counteracts the threat of premature ageing in the
adult soul. It also takes up the point of the patients' justifiable longing for rejuvenation and helps to prevent the retrograde soul development that may result from this. Steiner said that adults needed to
develop 'spiritual idealism'; he was referring to the need for the spirit
to release ideal aspirations from the depths of the sentient soul and
sentient body. In this way, we can 'be learners, learners from life,
all our lives.'[80] It is a way, however, that is followed in the light of
full awareness.

We do not stay young but always grow young over and over again
by drawing on the resources of the sentient body, the sentient soul
and the intellectual and mind soul; the powers we gain are taken
hold of by the awareness soul and devoted to the spirit. Goethe said
that people of genius went through 'repeated puberties.'[81]

Drawing on resources that go back to puberty (and before) may
however also lead to new crises in old age. Goethe was seventy-four
when he fell in love with nineteen-year-old Ulrike von Levetzow,
an affair that brought happiness and pain and finally renunciation,
and took Goethe back to the years of his puberty. He was able to
transform the youthful resources he had gained and give new life
to the wisdom of his old age that yielded not only the *Marienbad
Elegy* and the completion of part two of his *Faust* but many other
works as well. Childhood powers that have been retained and
transformed 'give the adult an independent creative mind.[82]

Developmental Disorders of the Awareness Soul

As already mentioned, the general human crisis that develops in the
late twenties will sometimes only come to awareness, or at least full
awareness, in the epoch of the awareness soul. If the crisis in the
late twenties is not overcome and the intellectual or mind soul does
not come to full development, the birth of the awareness soul will
be incomplete and the development of this soul impeded. The same
holds true with reference to underdevelopment of the sentient soul
and sentient body. The picture used to reflect the whole process in
which the different soul principles develop also indicated possible
consequences: If the leaves or calyx of a plant are not fully developed,

or if one removes both leaves and floral envelopes as it comes into flower, the flower is likely to wither. Such a withering process is also likely to occur in the life of the awareness soul when this is not provided with 'nourishment' by the other soul principles. In that case one will not see a life of the mind that proceeds in the light of awareness but merely the shadow cast by this light, i.e. intellectualism at the level of the awareness soul. This intellectualism may prove highly successful in the material world, but when it addresses itself to things of the mind and spirit these will become lifeless, frozen abstract concepts.

A different situation arises when the earlier soul principles have fully developed but now the step from intellectual or mind soul to awareness soul is not taken. Individuals who do no more than reflect, who ponder what has gone before but do not come to perceive the essence, the reality of the spirit, will find that the intellectual or mind soul, too, becomes increasingly more lifeless. In principle this is the same process as when the sentient soul wanted to hold on to things as they were. Continued reflection will go downhill and become fruitless brooding or else lead to everything being reduced to a scheme. The mind revolving upon itself becomes a comfortable habit and finally empty routine; inwardness threatens to become a cosy retreat. Bourgeois tendencies also threaten if things of the mind and spirit are cultivated during leisure periods and no effort is made to let them enter into everyday life. These individuals are too much caught up in themselves and do not progress to the point where new relationships with the world are established in full awareness. The inwardness of the intellectual and mind soul turns into egotism. The ideal virtue of the awareness soul, presence, is not achieved.

Many of us must in all honesty admit that we have this kind of problem, at least in part. In the case of the awareness soul we are even less able to say that we have been able to meet its demands in full than we were in the case of the intellectual or mind soul. This is partly due to the fact that the awareness soul is only in the early stages of development. During the sixth seven-year period the developmental disorders of the awareness soul lead to the mid-life crisis that one hears so much of nowadays.

Mid-Life Crisis

During the sixth seven-year period–sometimes around the age of thirty-four, sometimes at the time of the second moon node at about the age of thirty-six, and sometimes only in the early or middle forties–an inexplicable negative mood will arise gradually or all of a sudden, and this will be accompanied by restlessness or a feeling of exhaustion. Some people lose impetus at this point and sink into a depression that will soon combine with concern regarding physical health. This may go to extremes in hypochondria or in the kind of early failure in life that is becoming increasingly common now, the average age being about forty years.[83] Others escape into busyness or seek oblivion in intoxication. Behind it lies the 'silent terror'.[84] Coming to after alcohol or drug intoxication or after a day packed solid with activities, people then experience empty frustration and a growing isolation in spite of intense efforts to make contact. Until then they relied on their own abilities or on divine guidance; now everything is cast into doubt.

The existential question: 'If only I knew what I really want from life,'[85] keeps coming up again and again, and this is a sign that the awareness soul wants to come to birth. The new existential crisis that arises at mid-life essentially represents not merely the question as to the meaning of life, but as to the real, ascertainable value of life in both its past and future aspects. It is not our life in this world but the work we do in it that appears essentially meaningless unless it is seen to have some inner value that is also brought to bear in the reality of the outside world. Everything we do becomes the responsibility of our own ego, and this has now come really awake on earth.

We have shown that the gradual progress towards the centre of the soul reaches its goal in mid-life. This progress takes place in every individual, but if it does not go hand in hand with growing inwardness of soul, the process of gradual concentration becomes one of soul shrinkage. The goal of the progress is always the 'centre point', that is the ego; the crisis will however turn this into a final or zero point. Now it is important if–like Goethe's Faust–we can find the 'all' in the 'nothing'. Every human being has to pass through zero–some more so, some less, some in greater awareness than others. If the awareness soul is not guided by the ego to give the individual adequate support in this, hopelessness and despair may result.

Self realization, in the twenties achieved by sensitively feeling one's way and by analysis in thought, now becomes a vital necessity and is taken up with much more will by the awareness soul at this stage. Self awareness gained without progressing to better and more conscious self realization seems pointless. Individuals who are caught up in the crisis of the awareness soul will not make progress with the self awareness they have gained, for this will not take them on into the future. Earlier on we spoke of a review of life as it has been so far from the high performance peak reached in mid-life; in these cases it remains caught up in the past. Some will try and avoid any form of self awareness, only to find that it comes to meet them later on in life and in very painful ways. Others discover that they have so many faults and imperfections that they begin to ask if there is any point in going on with life.

This is where we encounter our own death. The experience of our own death in mid-life will not be a single or repeated event but forms part of the background throughout the crisis period. How many years are left? Am I going to make it in those few years, seeing that there are so many goals and ideals I did not achieve earlier on in life? Is there any point in going on at all, in making the effort, considering the fact that death is going to put an end to it all? Questions like these may in themselves be a step in the right direction and lead individuals to perceive their mission in life and to experience the ego as a spiritual principle that will not come to an end when they die. On the other hand such questions may lead to utter despair.

The encounter with one's own death in mid-life is further enhanced in so far as people now seek death more than they did earlier on in life. During the forties the 'number of suicides begins to rise sharply'. American statistics have shown that among women the incidence of suicides reaches a peak between the ages of forty and forty-five, whilst among men it continues to rise right into old age.[86] According to German statistics, twice as many men as women commit suicide. All in all, the incidence has so far 'risen from generation to generation and appears to be rising further.'[87]

The mortality rate due to illness also 'shows a sudden increase between thirty-five and thirty-nine.' Mozart, Raphael, Chopin, Rimbaud, Baudelaire and Watteau belong to this group. In the late forties the rate then comes to coincide with the average rate again.[88]

Some people develop mild or relatively severe mental conditions in mid-life. The German poet Moerike for example withdrew from the world and into hypochondria at the age of thirty-nine; he was less and less able to follow his profession as a clergyman and his creative powers had gone. The Swedish dramatist Strindberg was thirty-seven when mental symptoms came to full expression in delusional states, though these did not prevent him from having genuine spiritual experiences.[89] The German dramatist and poet Heinrich von Kleist felt that he had come to the end when he was thirty-four and shot himself. The Dutch painter Van Gogh did the same when he was thirty-eight, but his most beautiful paintings in luminous colours were painted during his last years when he was mentally ill. The Austrian composer Hugo Wolf started to go insane at the age of thirty-six; he died in an asylum when he was forty-three. The German composer Robert Schumann showed signs of profound exhaustion at thirty-four; his bouts of depression and physical ill health grew much worse at that point and he died at the age of forty-six. Both composers fought their disease to create their great works. The German philosopher and critic Friedrich Nietzsche began to show signs of insanity in his late thirties, and developed full-blown insanity at the age of forty-four. Steiner said that the constitution of Nietzsche's mind could only be understood in terms of the psychopathology, though this must not be allowed to determine the question as to where truth and error lie in Nietzsche's ideas. In his case, brilliance shone through a pathological medium.[90]

Friedrich Hoelderlin–Illness and Poetic Work

The life of the German poet Friedrich Hoelderlin vibrantly illustrates the working of the soul principles and their crises. This throws new light on the relationship between creative genius and mental illness.[91]

In the first two seven-year periods he lived a sheltered life in the gentle and delightful Neckar valley of Swabia (born 1770 in Lauffen on Neckar, later moved to Nuertingen on Neckar). Looking back on those days the poet later wrote:

> Mich erzog der Wohllaut
> Des säuselnden Hains
> Und lieben lernt' ich
> Unter den Blumen.
> Im Arme der Götter wuchs ich groß.

> The melodious sound
> Of whispering woodland leaves
> Filled my formative years.
> Learning to love among the flowers
> I was held close in the arms of the gods.

His ether body was greatly attuned to the world of nature and found it difficult to relate to the human beings around him. 'I understood the stillness of the ether. The words that people spoke I never understood,' he confessed in the above poem. His mother's strictness cast a shadow on his childhood; she decided that he was to be a clergyman. Growing up without a father he missed paternal authority and guidance. All this impeded the process of incarnation, though this was already destined to be incomplete because of the particular nature of an earlier Greek incarnation.

At the age of fourteen Hoelderlin was sent to a monastery school. A great struggle soon developed between a strong, passionate sentient body and the clerical walls of the monastery. He chose his own heroes and endeavoured to follow in their footsteps. At the age of eighteen years and seven months, the time of his first moon node, he went to Tuebingen to study theology. There, Hegel and Schelling were his friends and Schiller became his ideal. At twenty-one, the search for ideals in his newborn sentient soul came to fruition in his *Hymnen an die Ideale der Menschheit* (Hymns to Human Ideals); he sought to express in poetic form the ideas he felt deeply about, to unite with them in vision in the way it had once been in the days of Ancient Greece.

A stanza from his *Hymn an die Goettin der Harmonie* (Hymn to the Goddess of Harmony) may serve as an example:

> Komm o Sohn! der süßen Schöpfungsstunde
> Auserwählter, komm und liebe mich!
> Meine Küsse weihten dich zum Bunde,
> Hauchten Geist von meinem Geist in dich.

> Come, my son! Chosen in the hour
> Of sweet creation, come and be my love!
> My kisses hallowed thee for our union,
> Breathing spirit of my spirit into thee.

Here Hoelderlin was struggling against a general 'tendency to be abstract' of which he himself was aware at this time. His sentient responses were so exuberant that he could not achieve the final living embodiment of the idea; his emotional tone therefore seems to express lack of inner fulfilment at times. The schism between emotional and rational extremes in his nature that had developed in puberty and still had its repercussions, caused profound pain, and Hoelderlin himself spoke of being 'torn apart' by this pain. The fire in his soul would suddenly change to icy coldness. 'I shiver; I see winter all around me,' he wrote when his plans for the future had come to nothing at the age of twenty-five and he returned to his mother's house at Nuertingen, seeing his life's path taking him to the 'snug parsonage' he dreaded. His self was struggling to achieve wholeness and equilibrium in his soul; it appeared to withdraw from the withering life of his sentient soul. 'Gone now is the morning of life. The spring-time flowering of my heart is over,' Hoelderlin confessed at this time in his poem *An die Natur* (To Nature).

The first stanza of his poem *Diotima* is like a continuation of this:

Leuchtest du wie vormals nieder,
Goldner Tag! Und sprossen mir
Des Gesanges Blumen wieder
Lebensatmend auf zu dir?

Now thy luminance restored shines down,
O golden day! And will the flowers of song,
Breathing life, once more
Sprout forth from me to thee?

Hoelderlin was twenty-six when love for Susette Gontard, the wife of a banker in whose house he was employed as family tutor, made a new day dawn in his life. But that day did not merely have its former luminance restored; it had become a 'golden day', casting a glow that originated in the poet's youth into the outer court of his mature years, the glow of the sentient soul shining into the epoch of the intellectual or mind soul.

Out of a new experience of the period of Ancient Greece–with its reflection appearing in the fifth seven-year period–the Greek priestess Diotima, as he called Susette Gontard, stepped down to meet him. Descending to meet him, she, whom he first called 'messenger of heaven', 'woman of Greece', became his 'kindly heart', his 'fair heart'. His own heart, torn until then between the fire of enthusiasm and icy coldness, between hymnic elation and elegiac contemplation, now received and returned the warmth of love. Filled with this love he no longer invoked the ideal goddess; the person he loved stepped forward out of the Greek ether to which he was wont to ascend in his transports. Diotima had come down to meet him on earth, where he hoped to gain a firm footing at last.

Almost three years passed like this, then the banker Gontard asked the family tutor to leave his house. At the age of twenty-nine he went out into the world again to make a place for himself as a poet. Again he failed. Friends helped and let him stay with them, but his heart no longer had a home. His youthful heart and mind, having previously let him feel the wholeness of his essential self, broke with the pain. Soul and ego now began to depart from the 'house' of his body and the first signs of mental illness began to show. Susette was dying of tuberculosis when he returned to Germany from his last position as a family tutor in Bordeaux; he arrived in Nuertingen making confused speeches, distraught and unkempt.

The disposition to mental illness had already been there, but it needed the strokes of ill fortune that Hoelderlin suffered to make the condition develop fully in his thirty-third year (see the chapter entitled *Schizophrenia: The Illness in Relation to Biography*). States of excitement and even frenzy alternated with periods of enervation and depression. Incoherence of thoughts, a tendency to produce neologisms and the senseless repetition of words and movements developed later. The schism between soul and world deepened and the patient withdrew from the world, at the same time showing hypersensitivity in his reactions to anything that reached him from it. He established a facade of polite phrases which he used when visitors came and escaped into the name 'Scardanelli' that he chose for himself. This happened particularly in the 'terminal stage' which began when he was thirty-five. During the second half of his life Hoelderlin was looked after by his family. He inhabited a small tower by the river Neckar in Tuebingen. The river by which he was born and grew up, now flowed past the window of his final domicile, where his body resided for another thirty-six years.

The life of his intellectual or mind soul had withered, and preparation for and development of the awareness soul only took shadowy form in the patient's life on earth, and this came to expression in a 'demented struggle for conscious awareness' that was dreadful to see (Blankenburg[91]). In the process, he completely lost touch with the world ; his body, the basis for the life of the awareness soul, was increasingly abandoned by soul and ego. In mid-life Hoelderlin's schizophrenia led to the shadow existence of a terminal state where life on earth was concerned.

In his creative work as a poet, the fact that the higher essential principles had partly separated from the physical body gave rise to a new upswing in and a loosening of language, making it accessible to new experiences. This is a level where the struggle for conscious awareness bore fruit and the awareness soul in its separation from the body became creative. Eternity shone through in its creative activity and condensed to images of great visionary power.

The poems Hoelderlin wrote when the illness was in its early stages have influenced the development of lyrical poetry to this day and become food for the souls of many. It was Hellingrath who rediscovered them at the beginning of this century. He called them the 'heart, core and pinnacle of Hoelderlin's work', and stressed, quite rightly, that the new style of these poems did not reflect a break but rather organic development in his work. This development was triggered by the illness, but was not in itself pathological, as some psychiatrists would maintain. A good image of the real situation has been given by the psychiatrist and philosopher Karl Jaspers: A diseased oyster producing a pearl.

A new hymnic phase followed, and the emotionalism of the hymns written in his youth turned into a 'blessed sobriety'. The language now showed the 'tautness' Hellingrath spoke of, with the word the determining factor rather than the thought. The human individual was undergoing a process of excarnation through the illness, but at the same time there was an incarnation of spiritual experiences at the poetic level that culminated in encounters with Christ. The human individual went through regression into a second, pathological puberty; the poet experienced a different kind of second puberty. Instead of maturity for life on earth at the level of the physical body he came to a cosmic maturity that gave rise to new language forms for the earth. Many things the poet and human individual had striven for out of his first puberty now came to fruition at that level. A few lines from the hymn *Patmos* may serve as an example, particularly if one compares them with the above stanza from a hymn written in his youth:

Es liebte der Gewittertragende die Einfalt
Des Jüngers, und es sahe der achtsame Mann
Das Angesicht des Gottes genau,
Da beim Geheimnisse des Weinstocks sie
Zusammensaßen zu der Stunde des Gastmahls,

Und in der großen Seele, ruhigahnend, den Tod
Aussprach der Herr und die letzte Liebe.

The bearer of the storm loved the simple mind
Of his disciple, and the watchful one
Clearly saw the countenance of the god
As they sat and shared the mystery of the vine
In the hour of the feast,
And in his great soul, calmly foreseeing,
Death uttered the Lord and final love.

Christ is the 'bearer of the storm' around whom the disciples are gathering for the Last Supper. The poet describes the event as if it were re-enacted before his eyes. The six 'A' sounds in the first two lines of the German original incarnate into language the rapt and attentive simplicity of John the disciple, for the poet is seeing the event through his eyes. At the same time, however, the disciple sees 'clearly'. A great calm breath takes us from the first line all the way through to 'final love', and in the German this involves an inversion where the words *und die letzte Liebe* do not follow *Tod* immediately but three words later. This puts an emphasis on *Liebe* that is not expressed in words, as the poet would have done earlier, but comes to direct expression in the way the words are positioned.

Later, however, regression due to the illness also had its effect on his poetic work. After a period of dissolution as regards both language and thought, a kind of 'terminal state' was also reached in the poetry; again it may be experienced as a repetition of childhood but at a pathological level. There were no further hymns. The poet put his impressions of nature in poems, but now let line follow line in a simple, childlike way; the rhyme was often 'stiff', the images tended to be 'stereotypes' (Hellingrath). Yet now and then the light of his mind and spirit, so far away now, shone through, with the objects of this world appearing as the 'reflected light' of heaven.

Der Erde Freuden, Freundlichkeit und Güter,
Der Garten, Baum, der Weinberg mit dem Hüter,
Sie scheinen wie ein Widerglanz des Himmels,
Gewähret von dem Geist den Söhnen des Gewimmels.

The joys, beneficence and bounty of the earth
The garden, tree, the vineyard and its keeper,
Seem a reflection of the radiance of heaven
Granted by the spirit to the sons of the teeming throng.

One particular biography has been considered in some detail to present a living picture of how crises in life may lead to the outbreak of an illness that may sometimes have its roots in an earlier life. (With reference to this last aspect, see the chapter on Hoelderlin in the author's *Der schizophrene Prozess*; this work also goes into more detailed consideration of the meaning of schizophrenia.[91])

Patients suffering from mental illnesses can wrest positive gains from their condition. Struggling with the disease they are able to gain new powers and abilities; in Hoelderlin's case these bore fruit in the sphere of art, for others they may do so in the sphere of human relations, though it is also possible that they will not come to fruition in the present but only in a future incarnation. The importance of this goes beyond the patient's own life. 'Hoelderlin's struggle against his illness can give courage and strength to others who are suffering from mental illnesses, helping them to come to terms with their own condition. But over and above this, Hoelderlin's poetic works can be a sign of hope for all of us, hope that the human spirit can overcome physical and mental illness.'[92]

The Ego between Body and Spirit

The genesis of the awareness soul, with the ego finding itself and being able to establish a conscious relationship to the spiritual world, pointed to the significance of the physical body, for the awareness soul arose through transformation of its powers. Anything that interferes with physical development in the first seven-year period of life creates obstacles for the genesis and development of the awareness soul in the sixth seven-year period. The lack of contact we have spoken of, and the isolation experienced in mid-life and from then onward, may have their roots here, and this is also evident in the histories of people who develop depression in mid-life. (Negative factors are generally found to have arisen particularly in the first seven-year period.) The isolation arises because these individuals are 'uncommunicative' and 'cannot enter into free and open intercourse with everything that reaches them from the outside world.' This type of character developed when the ego found the physical body hardened when it came to the transformation of physical powers in the thirties.[55]

Negative psychological factors during the first seven-year period come to mind, particularly lack of and–more rarely–excess of warmth in the family nest. Due to external factors, the soul does not fully swing in harmony, and this is passed on to the body with the result that the soul forces are not sufficiently involved in building the body and it becomes too physical, which will cause premature hardening. But there are also physical factors: Vaccines used to prevent any kind of febrile infectious disease in childhood may cause hardening, as do high doses of calcium in the prevention of rickets.[93]

In people who are uncommunicative, the ego gets stuck in the developing awareness soul and does not make adequate contact with the world. This is a repetition at soul level of what happened during the first seven-year period when the individual's essential nature was not able to penetrate the living body. Once again, and this time due to factors in the first seven-year period, we come up against a pathological situation in mid-life that takes the form of underdevelopment of the awareness soul. The ego, situated between body and spirit, does not, or only partly, succeed in taking soul development up to the realm of the spirit: The developmental curve then comes closer and closer to the downward curve of biological development.

At this point the symptoms of the mid-life crisis take concrete form: Suppression of the soul's efforts to reach the spirit and bondage to the physical body are experienced as a lowness of spirit that may turn into a full-blown depression. Any depression that occurs after mid-life essentially has its base in such a pathological relationship to one's own physical body. (The fact that this tends to have its focus in the liver will be discussed at a later stage.)

Bondage to the physical body also comes to expression with this type of depression in that there is a marked element of hypochondria. Being caught up in the body in such a pathological way, sufferers also have intense awareness of the gradual increase in physical debility, so that physical symptoms that normally remain subliminal come to awareness and arouse concern. Hypochondria is not simply something elderly people are apt to develop; it arises because they are not growing old in a healthy way.

Psychiatric and Nervous Diseases of Old Age
Like the hypochondria and depression of old age, all other pathological changes coming to expression in mental and emotional symptoms

at that time have their root in the process that begins in mid-life. The delusion of poverty, for instance, that sometimes comes with an advanced age arises from unconscious participation in the physical decline. If the soul is so much aware of this process that it is simply full of it, this will also determine the way one's financial situation is regarded. These old people are afraid they may, and indeed believe they do, fall into poverty–financially and in terms of clothing and other objects. From there it is just one step further to the delusion of having been robbed: The missing money or clothes have been stolen. Both kinds of delusion base on the genuine physical 'impoverishment' that occurs in old age, but the old people are stealing from themselves, from their own physical bodies. Ageing means that the astral body is feeding on the ether body, and later the ether body on the physical body.[94] This characterization given by Steiner reveals another developmental component. On the one hand the physical body releases soul development, on the other the higher principles are feeding on its powers and making them serve the life of the spirit. The body is conscious of depletion because of this, a sensation that plays a role in many illnesses and in many experiences people have at a more advanced age.

The consequences of this for the physical body are partly physiological, i.e. regression or involution is to some extent normal, but there can also be morbid sequelae at the physical level. These, too, arise because soul and ego are too closely bound up with the physical body. This pathological bond leads to enhancement of the physical decline, so that it overshoots the goal. It is well known that worrying about one's health is harmful to health, though this is rather a superficial way of putting it. The physical problems arise because the essential principles behave in a way that makes individuals feel concern for their physical health. There are two aspects to this: In the first place, ether body, astral body and ego partly let go of their living relationship to the physical body, and this is something that always happens as we get older. On the other hand these higher principles may connect with the body again and establish secondary bonds. This is a pathological process that due to its very nature cannot lead to a new, life-filled incarnation, for that belongs to a future life on earth. The new bond with the physical body occurs at the level of a nervous system that itself has largely grown dead at this point and can no longer mediate formative elements but merely conscious awareness, i.e. awareness of physical decline. This in turn will further deplete the life of the body; physiological involution in the nervous system is pathologically enhanced and spreads to the rest of the organism. Neurasthenic conditions in the widest sense of the word are the consequence.[95]

This situation, i.e. the physical consequences of an ageing process they have not been able to master, is responsible for the increase in degenerative diseases among the elderly, and this may combine with a tendency to proliferation and tumour growth (only brief reference can be made to this). Particular emphasis must be given to the increase in degenerative diseases of the nervous system that develop above all in the second half of life. Atrophy of the brain due to increasing cerebral sclerosis or of its own accord is an example. Multiple sclerosis, a condition that may even develop the first half of life, has a degenerative and an inflammatory component. (Inflammation essentially belongs more to youth in humans.[95])

We have shown that the awareness soul comes to life from the head pole. We can therefore regard nervous diseases in the second half of life and degenerative conditions altogether as shadows that reflect the development of the awareness soul at the physical level. When the brain atrophies in old age it no longer receives sufficient life from the blood; in the same way the underdeveloped awareness soul is not receiving sufficient life from the heart, where it has its roots. The birth pangs of the powers of heart are also the birth pangs of the awareness soul. They come to physical expression in the neurocirculatory asthenia we have discussed, for as we have seen, this belongs not only to the twenties but also to mid-life.

Failure of the brain in old age, which under certain circumstances can be traced to inadequate or disturbed physical development during the first seven-year period, casts its shadow on all the mental conditions seen in geriatric patients, for these are metamorphoses of the mental diseases we will be discussing later on (and others). The delusions one sees in old age in particular are complicated by poor comprehension and judgement, forgetfulness, and sometimes also poor orientation in space and time. Objects are misplaced or lost and old people think they have been stolen. All the above symptoms are due to reduced brain function. On the other hand the instability of feelings, emotional excitability and irritability noted in the elderly also have to do with the brain. It is through the brain that the ego becomes aware of the stirrings in its own soul, so that it may try and control them. If the organ for such perception is not functioning properly, the ego can no longer intervene the way it did before and the impulses that surge up in emotional life can no longer be controlled.

The problems this creates in social life are well known. The isolation of the elderly no doubt has its external causes, for people withdraw from them. But as we have already seen, isolation can also arise from within. Ageing people who become too much bound up with the body loosen their connections with the people around them. When interest focuses mainly on one's own existence as determined by the body, others soon cease to take much interest.

The consequences of wrong development due to underdevelopment of the awareness soul even affect one's philosophy of life. Sometimes the mental and intellectual consequences of development having come to a halt in the first half of life only emerge at this point, or they may be given dramatic emphasis: Atheism and materialism come to painful experience. The more the soul is caught up in the body, the less potential is there for perceiving the working of the spirit in the world. If one's thinking is tied up with progressive involution in the brain, death as the end of that process must inevitably appear to be the ultimate thing. If such a philosophy of life is projected on to the world, one gets the atheistic view of materialism and in the end this is no longer a mere theory in one's head but becomes a compelling principle for the conduct of life.

Therapeutic Aspects
Initially one will again have to consider earlier disorders in soul development. If individuals have lost the capacity for a living sentient response, or if their sentience has regressed under the light of the awareness soul, treatment has to start at this point. If they find it difficult to think clearly and develop inwardness of heart and mind, they will need to catch up in these areas.

Yet neither they nor their helpers should simply recapture the style of the sentient soul or of the intellectual or mind soul. The awareness soul may be underdeveloped, but due regard should always be paid to its style, i.e. the style of that particular stage in life. When children have much catching up to do, yet are of an age that would put them in the eighth year at school, it would be far from ideal to put them in a sixth-year class. The right way is to provide extra coaching for them whilst keeping them in the eighth-year class. The same applies to any form of catching up, but above all to any catching up that has to be done at the awareness soul stage, for usually a higher level of awareness exists already and has to be taken into account. This will of course be a major problem in many cases, but helpers should always keep it in mind. The individuals concerned often know quite well themselves that they need to be addressed in terms of the goal of their development.

If the soul principles that precede the awareness soul are well developed, or if delay in development has been made up, the indefinable longings felt earlier and the desire to find the spirit by using the faculty of thought give way to a marked desire to gain full understanding of the self and of the world, and this should be nurtured. The question as to where in life so far elements can be found that hold significance and true value for the world around one will be considered together with the helper. Reviews of past experience and the reflective review of life may now develop into a review of essential things. Looking back over the day the question may be put: 'What has had real significance today? Can I find anything that was of general validity and had spiritual qualities?' In the morning a will element may be brought in and the resolution made to consider the true significance of certain things and make them more significant during the day that lies ahead. Such resolutions may also be practised in the twenties to prepare for the awareness soul.

Steiner advised the practice of a nightly review of the day. This is a fairly advanced exercise in preparation for spiritual training and not everyone is able to cope with this immediately. The exercise is of particular value for 'strength gained in forming mental images'. One achieves it by reviewing the events of the day in reverse, at the same time attempting to see oneself in the picture. Such an activation of the faculty, basing itself not on the outer progress of the day but on personal will impulses, also helps to gain distance from the personal experiences of the day. It is important not to dwell on individual events as the day is reviewed, nor to reflect on them etc., but to let them pass review as though they related to someone else.[56]

All such exercises address the will of the awareness soul; as it gains in strength we develop the ability to think ahead and this helps to make the transition from the intellectual or mind soul to the awareness soul. It is also possible to practise projective thinking as such; this will activate the underdeveloped awareness soul. One may follow the process of a plant growing from seed in the mind's eye, for example; this is the first part of the 'seed' exercise given by Rudolf Steiner.[96] Anything read should no longer merely be reviewed and reported, but freely re-created in one's mind, without going into too much detail. This leads from reflection to projective thinking, and the original may of course be consulted to see if the re-creation has been accurate.

Starting to do such exercises even whilst gaining the necessary self-awareness, one is then able to help the other person, or oneself, not to get caught up in the past or in personal concerns. The future comes into it when out of personal realization the question arises as to the special mission that is given to every individual. And a further question, 'How can I find better ways of fulfilling my mission? How can I gain new strength to do so?' establishes the link to objective spiritual realities.

This is the time when the will can become really active in the soul, and personal experience through exercises achieves vital significance. The awareness soul is not trained by knowledge and understanding, for that is its life element. It is trained through the reverent devotion we have already mentioned; this establishes a connection with the source spring of the life of the awareness soul, which is the living experience of the heart. Active devotion, with one's thinking, feeling and will activity coming together at the centre of the soul, needs practice, however.

Reverent devotion leads to meditative mood and meditative exercise of the awareness soul, for the meditative element is the vital core of this soul. The way from the world to the centre of the soul is followed again, this time entirely determined by the will. Individuals doing the exercises hold the mood of reverent devotion and consciously and deliberately shut out the world. A mental image is placed at the centre of the soul and one concentrates on it, beholding it and experiencing the sentient response. To prevent a compulsive element entering in, will power is used to 'wipe out the image' at intervals. According to Steiner, symbolic images are more effective than images taken from the outside world. Even the images we have of light, of wisdom, of warmth and of love will achieve something. Even more will be achieved if the image of warmth as the symbol for love is created and meditated. Such images serve as 'educators of the soul,'[27] i.e. they represent not only steps in training but help in the crisis of the awareness soul.

The way to the innermost soul, where the ego can find itself and at the same time also access to the world of the spirit, finally leads back to the outside world. We become aware of gaining a deeper relationship to it than we had before we undertook such exercises or meditative reading. We gain living awareness of the essential realities of the world and of our own lives. A ray from the world of the spirit has been mediated by the ego and come to be our companion as we step out into the world, and we will then be able to develop a new and more deep-rooted competence in the world.

Therapy for soul and spirit is now more than ever a form of self-help and self-healing, but medical treatment and remedial eurythmy will also give valuable support. Both are of major importance where the developmental disorder has in turn affected the physical body and caused functional or organic disorders. In these cases one will primarily consider the organ or organs involved, but gold therapy is recommended to support the awareness soul via the heart. This has already been discussed. Lead in relatively high potency acts on the awareness soul via the nervous system. Gold helps to establish a new relationship to the world of the spirit and it helps to do so out of the centre. Lead supports the awareness soul particularly in its efforts to gain understanding. Low potencies of lead enhance processes of degradation, and this is sometimes necessary; in higher potencies the metal intercepts an overshooting catabolism and counteracts sclerosis and degenerative processes.

In remedial eurythmy, the *Liebe-E* (E of Love, with the E pronounced like the first part of the ai in pain) exercise that relates to gold is suitable and the 'T' gesture. The first part of this, raising the arms, establishes a relationship to the environment and to the cosmos; the second part, where the arms and hands are brought down to the head, establishes a connection to the head and from head to heart. The 'U' (pronounded oo) gesture picks up the thread of lead therapy. If the hands are put together and moved downwards from above, this particularly relates the conscious ego to the earth. This addresses the human being who is standing securely on the earth. Raising the arms again in the 'U' gesture, the individual then turns his conscious mind to the spirit again.

Much can also be done to prevent physical and mental illness in old age. Apart from a lifestyle that is as rhythmical as possible and orientated towards the spirit, much can be done medically and with eurythmy and other forms of art therapy. Turning to the spirit does not mean turning away from the body and leaving it to itself and to gradual decline. A valuable instrument like this needs to be properly cared for even in old age. On the other hand this should not become an end in itself but it should help the ego, so that it can use the instrument to let the melody of the spirit grow purer and purer in sound.

Chapter Four
Neurosis and Psychosis in the Developmental Context

During the third seven-year period, when the astral body achieves independence, we encountered the type of angioneuropathy that was shown to be due to the astral body becoming excessively or inadequately bound up with the physical body. This neurosis produced both physical and mental and emotional symptoms. An addictive desire that was initially felt in the souls of those young people later entered into the physical sphere. In the fourth and also towards the end of the third seven-year period we encountered a 'vacuum neurosis' the symptoms of which were largely mental and emotional but then again also related to physical symptoms. Finally, the neurocirculatory asthenia that is characteristic of the sixth seven-year period began with a history of emotional disturbance and ended with organic disease of neurotic origin that in turn involved further mental and emotional symptoms.

The physical and mental cum emotional symptoms of neurosis (and later psychosis) can be confusing. It may be helpful to consider the two directions the disease process may take, as different symptoms belong to one or the other. One direction is from soul to body, the other from body to soul. Using this approach we are no longer tied to a fixed notion of the disease as a defect in either body or soul. Instead, we enter into a living developmental process, the disease process rather than the disease, and this moves between the physical and emotional planes, showing itself sometimes more at one level and sometimes more at the other.

Physical Illness Originating in the Soul

The connection between the two planes is particularly evident in cases of 'organ neurosis'. We have already considered neurocirculatory asthenia, which is probably the most important form of organ neurosis. This does not affect the physical heart the way a myocardial infarction, myocarditis or valvulitis does, but cardiac function, i.e. the movement of the heart and its blood supply, is abnormal. In the same way nervous dyspepsia, a neurosis of the stomach, means that secretory function and hence the gastric juice is abnormal. The characteristic feature is always abnormality in the way the blood and the juices flow and in the movement of an organ.

This brings us to the water organism of the body which serves to maintain organic functions and thus life itself. The ether body is particularly active in this; it builds and shapes the physical out of the watery element and at the same time ensures that the airy principle of internal respiration reaches the whole organism via the watery element. The astral body is primarily active in the air organism, and the ether body ensures that this activity reaches the physical body via the airy principle.[1] This indicates that a living relationship exists between body and soul. In the view that anthroposophy takes of man, the whole body is ensouled. If we limit the soul to the brain—present-day theories maintain that it is via the brain that soul and body are related—we also set limits to a living comprehension of the relationship and instead have an abstract, mechanistic circuit system connecting body and soul computer-style.

Emotional experiences that the soul is unable to assimilate act primarily via the soul aspect that is active in the body, i.e. the sentient body (or astral body), influencing internal respiration, i.e. the exchange of oxygen and carbon dioxide in the organism. This is the point where the ether body that is active in the water organism takes in the mental or emotional factor. This leads to changes in blood circulation, in the juices and in the movement of organs. If the disruption essentially affects the organism as a whole, neurocirculatory asthenia develops as excessive or inadequate astral involvement is transferred to the physical body in the way we have already described. If the disorder concentrates on a particular organ system, the diagnosis is 'organ neurosis'.

The disorder may however go deeper and affect the organ itself, i.e.

its solid substance. The organ is then 'deformed' by one of the degenerative diseases we have discussed or by inflammatory changes. 'Astral activity is getting too much of a hold' on the organ.[2] Organic disease may be the direct outcome, but there may also be symptoms of an organic neurosis that indicate that such a condition is developing. Heart burn and a sensation of fullness in the stomach may indicate a neurotic functional disorder, but they may also be the first signs that a gastric ulcer is developing, in which case the heartburn tends to progress to become a burning or boring pain. The psychological changes in the patients' prior histories are similar to those noted with gastric neurosis. An infantile attitude to life may combine with a 'burning ambition' that could not be fulfilled and therefore affected the stomach.[3] The burning felt in the soul is changed into physical pain and physical illness, and finally destructive processes caused by soul forces coming to awareness and being suppressed and pushed down into the body cause physical substance to be lost through inflammation as a gastric ulcer develops.

In this way the wrong mental attitude can lead to physical illness. However, this should not simply be to be the consequence of what went before but also as an opportunity for the future. Gastric symptoms and the pain of a gastric ulcer can also make patients realize that there is need for a change of some kind in their lives. Once they have grown more mature in their souls and are more able to be aware of their ambitions and to cope with them, they will also be better protected against the recurrences of the stomach complaints that are liable to develop.

Other factors in the psychological history tend to have similar effects. Febrile illnesses of known or unknown origin are often preceded by a state where one was 'too much given up to the outside world', a kind of 'fever in the soul'; this can combine with other factors (a chill, or an infection) and cause a fever.[4] Mistakes made in the upbringing and education of children may be responsible for physical illness in adulthood. In all these cases the direction is from soul to body, though in the latter case the problem arose from outside. Psychologically it may lead to behaviour disorders in childhood, and physical illness may follow later.[5]

It is characteristic that these are not psychological conditions that become physical diseases, but wrong mental attitudes or disruptive influences from the environment out of which the physical body, working together with the ether body, fashions physical diseases.

Psychosomatic medicine now recognizes that there are many instances where this connection exists, and further psychosomatic conditions will no doubt be discovered. It is now generally recognized that mental and emotional factors play a role in every physical illness. If the effects of earlier earth lives are also taken into account it will be evident that in that case, too, the direction is from soul to body. A hereditary disposition may be enhanced by negative elements from earlier life. Illnesses that arise through accidental injury, i.e. an external event, may also have their causes in the soul life of an earlier incarnation; the ego seeks them out in the biography to help it shape the individual destiny. From this point of view every physical illness has a psychological history.

Physical illness may thus be considered the end stage in a process out of which new attitudes may develop, and this takes us back again to the level of the soul. On the other hand it may provide the basis for further illness, and this will again be in the sphere of the soul.

Psychological Conditions Arising from Physical Causes

Many people have known occasions when they woke up in the morning to find that soon after waking a strangely negative mood and a nervousness developed for which there was no immediate explanation. It is possible that something had been a worry to them some days or weeks ago, but they thought they had overcome it. If we follow this up we find that something has not just gone to the stomach of these people but to their livers, and that the negative mood arises from this, sometimes in conjunction with a slight sensation of pressure or fullness. The mood and the nervousness will soon pass, only to return when something else happens. People will then ask why this particular person is always nervous and so easily upset. The answer tends to be that they are neurotics. It may also be said that they have a psychoneurosis that may be due to negative experiences in childhood. Such experiences–if they exist–and recent worries that have not been fully overcome have however affected the liver; functional disorders affecting the flow of bile and the circulation of the blood have been caused in the liver by the ether body. The dynamics of this physical condition are then conveyed to the soul by the ether body. Blockages, spasms and explosive discharges in the movement of the blood, the juices, and of organs generally, become blockages, spasms or explosions in a neurotic soul life that then reacts with negative moods, nervousness and aggressivity towards the world.

Organ neuroses may become more deep-rooted, and so may disorders of the soul. Negative moods turn into severe and hopeless depression, nervousness into a paralyzing fear of life that makes any kind of activity difficult if not impossible. At this point the severity of the presenting condition is out of proportion to the painful incidents of recent origin, and in fact these need not even exist. (There will however always have been such events in the past.) The physical history sometimes includes organic liver disease, or it will be found, among other things, that liver metabolism is abnormal during the depression, and that this abnormality persists for some time and involves the internal function of liver cells. Endogenous depression, the commonest psychosis involving the liver, will be diagnosed. We shall consider this further when we come to psychiatric conditions in general.

Psychoneurosis thus develops out of milder, functional disorders in an organ system; in psychosis the organ is more profoundly affected. In view of the above we may say that psychosis is a psychological condition with an organic base where 'deformation' at the organic level is transferred from the physical body to soul life by the ether body. Any physical predisposition points to events in an earlier life in this case.

According to Steiner, psychological illnesses are always due to the soul and spirit principle taking on the structure of the physical and etheric principle, so that it is cast in the physical or etheric mould.[6] Elsewhere Steiner spoke of 'psychological illnesses' where the principle of gravity that related to the physical body entered into the life of the soul.[7] Melancholia may be experienced as the immediate consequence of this; it is a form of depression where heart and mind grow heavy.

This characterization and the fact that Steiner spoke of 'so-called' psychological illnesses reveal the difference between psychological and physical illness. The history of physical illnesses generally shows no psychological illness but wrong attitudes on the part of the sufferer or others and incidents that have not been fully coped with. The history of psychological illness always shows a more or less serious functional or organic condition affecting a particular organ system. In the first case the soul life is involved in the process in which these diseases evolve in the body; in the case of mental illness the disease is transplanted from the physical to the mental sphere. One therefore cannot speak of illness in the soul the way one does of illness affecting the body. 'Actual diseases of the spirit or soul are an impossibility.'[8] These diseases do not arise out of the essence of the soul or spirit.

On the other hand Steiner spoke of a process leading to illness that consisted in a physical principle alien to the soul being transferred to soul life. The soul does not create illness out of itself, but it has sickness put on it by the body. It grows ill not in its essential nature but in its life, its life substance, i.e. its thinking, feeling and will activity, the three fundamental powers of soul life that relate closely to the body (cf. Steiner's concept of the three aspects of the human organism). Steiner therefore spoke of an 'abnormal soul life'[5] or 'pathological soul life'[9], of 'defects in the will' and 'defects in thinking' with 'deformation of thoughts',[10] all of them symptoms that have their roots in the body. The term 'psychological condition' therefore always refers to illness in the life of the soul in this context. (For mental illness, or insanity, as distinct from what we here call a psychological condition, see under *The Spirit–Illness and Healing*.)

Connections between Physical and Psychological Conditions

Something physical and psychological conditions have in common is that the disorder in the area where the roots lie has abated or disappeared by the time the pathological condition shows itself. This applies to psychological problems preceding physical illness and just as much to the physical problems preceding psychological illness which will then have entered a less acute and more chronic stage. That is why Steiner spoke of 'subtle' physical disorders that were at the root of psychological conditions.[11] This is due to the fact that soul and ego partly withdraw from the physical sphere, just as physical illness arises when soul forces come in too deeply. 'Insanity arises when the spirit itself withdraws completely.'[12] Added to this is the fact that ego and soul are 'pushed out' by the disease.[5]

Ego intervention in the disease process may be reduced once physical conditions have become chronic, and this means that the potential for self-healing is also reduced.[13] The image of 'organic resignation' presents itself; it is the opposite of the 'organic rebellion' we spoke of earlier. This organic resignation may also combine with 'mental and emotional resignation'; ego and soul withdraw not only from the body but also from the world to which the body is the connection. It is possible to observe this in the preliminary stages of some psychological conditions, and it may grow more marked as the illness progresses. At the same time ego and soul do not come free of the body and the world. They are unable to withdraw completely–this happens only when death occurs–and are therefore more or less exposed to the body and the world because they have grown remote or are no longer in tune and therefore do not come to terms with them. A physical condition that has not been overcome and where the active link with ego and soul was loosened then takes the form of a psychological condition that affects both ego and soul.

Resignation of ego and soul combines with a specific disposition in psychological illness, and this disposition also play a role in the relationship between soul and body. A soul either is or is not predisposed to depression not only in relation to the world but also where the effects the body has on it are concerned. This alone explains why relatively minor abnormalities in liver metabolism lead to serious illness in the sphere of the soul in some subjects but not in others. Conventional scientists therefore feel that they cannot accept the specificity of those minor metabolic abnormalities; in view of the fact that they also occur in normal subjects such changes are not considered specific for psychological illnesses. In fact they can only be given their rightful significance if the concept of disposition is extended (see the chapter on depression).

Another objection, one that can actually take us further, concerns the quantitative aspect of the physical disorder. How can such minor abnormalities lead to such a severe psychological syndrome? This provides a good illustration of the fact that thinking in terms of quantity and sheer matter must become thinking in terms of quality and processes if one wants to do justice to diseases as living processes. Research into hormones and trace elements has shown that very small amounts of physical substances produce qualitative effects in the sphere of the soul. Subtle rather than massive physical changes have differentiated and profound effects on soul life. A severe hepatitis (inflammation of the liver) characteristically does not go hand in hand with endogenous depression; it is only when the inflammation has turned into a more subtle disorder that this type of depression may arise. Massive endogenous depression does not normally combine with hepatitis; it may actually improve and disappear completely if hepatitis develops.

This has to do with the relationship between the constituent principles in psychological and physical illnesses that was discussed above. In severe psychological conditions, the ego, soul and sometimes also the ether body have withdrawn from the physical body and produced their own type of syndrome. In the case of serious physical illness ego, soul and ether body are fully engaged in the physical body and no longer able to produce a differentiated syndrome of their own. They are merely able to react to the physical condition in a nervous, depressive, excited or similar manner, depending on the individual disposition.

These and other kinds of psychological symptoms are therefore not specific to the affected organ or organic system in a case of serious physical illness. Any hallucinations or delusions that may develop do not, or not entirely, arise from the affected organic system and do not represent an organ-specific psychosis. The latter can only develop once the physical symptoms have gone down. (A symptom not usually seen with endogenous psychoses is clouding of consciousness, a condition due to the brain being involved in the physical disease process.).

The following image presents itself: A psychological illness is like a plant that has only very fine roots in the ground, with most of its development in air space above ground. Yet if we want to get rid of the plant the roots must be taken out as well; otherwise it will keep coming up again. In the same way we must go to the root of any psychological condition we wish to cure, i.e. treat the body, the organ, even if the roots are quite tenuous and the illness has developed mainly in the sphere of the soul.

The above facts and considerations helps us to understand Steiner's first maxim for the anthroposophical approach to psychiatry: 'In socalled "psychological" illnesses the primary factor is in fact to be found in the organ systems.' The treatment of these illnesses, he said, should be medical and it should be based on the organs.[14] This applies not only to psychoses but also to neuroses, and the greater role that environmental factors play in the latter should not lead us to think that in this case the basic pathology and basic treatment do not relate to organ systems. Any kind of abnormal soul life has its base in abnormal organic life.[5] We therefore cannot be onesided and treat only soul and spirit.[15] Spiritual science must most emphatically state that so-called mental or psychological illnesses have to be traced all the way to the organology of the individual.'[5]

Diseases to Hinder or Help Development

We have established that the two major trends in soul development are incarnation and excarnation. Having united with the body the higher principles rise above it again in stages, and the body then becomes a kind of instrument for their further life and development. The two directions in which illness arises, and that give rise to organ neurosis and physical illness on the one hand and psychoneurosis and psychosis on the other, were either into the body or out of the body. The opposite direction always makes itself felt as well, as in the biography: Every physical illness also involves the soul, and every psychosis or neurosis the body. The emphasis will however be either on incarnation or on excarnation, going into the body or coming out of it.

The degenerative and inflammatory conditions we have so far discussed may thus be seen to relate to an incarnating process that has gone wrong and become pathological. Mental and emotional powers take up the style of the nervous system and attach themselves too strongly to the body; degenerative diseases with destructive elements that are characteristic of the nervous system develop. Or, these powers assume the style of the metabolic system and enter too deeply into the body; inflammatory diseases involving dissolution similar to the digestive processes develop. We noted that this was preceded by a developmental disorder. The powers of the soul did not separate sufficiently from the body, they were rejected by the environment, or they gave themselves too freely to the world in a 'fever of the soul'. In each case this has physical repercussions. If the powers have not come sufficiently free from the body, one is usually dealing with a primary developmental disorder of the soul, in the other two instances with secondary developmental disorders due to those physical repercussions. In each case an earlier stage in the incarnating process makes itself felt in more intensive bonding to the body. It means, however, that incomplete incarnation may now make further progress and patients will appear better incarnated after their illness than they were before. A new developmental aspect shows itself.

The meaning of physical illness has already been considered. The consequences of wrong mental attitudes became apparent in physical illness and new attitudes could be developed (e.g. with a gastric ulcer). From the developmental point of view we may say that partial retrograde development to an earlier stage in life will not only mean catching up on an imperfect incarnation process but also the possibility of a new start from the earlier level. People who lie there helplessly in a high fever or suffering from a serious inflammatory condition may be said to have returned to a condition they knew in childhood, when they were dependent on others. Clearly it is possible to make a much more thorough new start by first going back to the stage where wrong development was in preparation or perhaps had already started. Physical illness thus presents itself as a partial, abridged and pathological repetition of incarnation in the body that at the same time offers an opportunity to grow up in a new way in some respect or other as recovery is made.[16]

The same principle applies in cases of psychological illness. The difference is that whilst the tendency is towards 'birth' with physical illness, the accent is on a movement towards death in the excarnating process of psychological illness. Partial withdrawal of soul and ego from the body creates the impression of a psychological dying process in psychoses; in the initial stages patients may experience this for themselves. Recovery, the first steps towards new development, does not initially lead out of the body (as in the case of physical illness) but into the body; after recovery the individual then learns to grow free of the body in a new and healthy way.

There is a connection between the two types of illness in so far as the pathological incarnation occurring with physical illness may suddenly change into the pathological excarnation one sees with psychological illness. The pathological incarnation of physical illness links up with the inadequate incarnation we have spoken of, and the withdrawal from the body of pathological excarnation in cases of psychological illness connects with the frustrated efforts to come more free of the body. Both are pathological variations of normal processes and can lead to continuation and enhancement of normal processes that have come to a standstill. This has already been mentioned with reference to physical illness. Examples relating to psychological illness will be given in the discussion of psychiatric conditions.

Summary

Different systems in the organism may undergo pathological changes, and different psychological conditions can develop on the basis of these systems. The question as to which system relates to a particular psychological condition leads on to another question, one that has already been broached, namely the relationship between soul life and physical body altogether. The threefold aspect of the human organism provides a systematic basis that will permit a first answer. The neurosensory system which has its centre in the head serves our thinking. The rhythmical system, which has its centre in the chest, serves our feeling life. The metabolism and locomotor system, which have their centre in the abdomen, serve the will impulses of the soul.[17]

Steiner refined this further by considering that in the lower organism the kidneys and bladder on the one hand and the liver and biliary system on the other related to different soul powers and functions, and so did the lungs and the heart in the middle region.[14] Basing himself on this he asked that psychiatry should recognize the psychic aspects of organs;[18] the state of soul should be assessed to determine 'where the fault may lie in the organism.'[19] 'Symptoms like these permit a much more definite diagnosis to be made than the diagnostic methods that are so widely used today.'[18] Certain psychological symptoms of psychological illnesses thus serve as pointers to abnormalities in a particular organic system, and as far as possible such abnormalities should then be looked for on physical examination. Treatment should certainly first of all be addressed to such organic abnormalities.

The relationship between soul life and physical organs changes in the course of development. In childhood, one particular soul power is engaged in building up a particular organic system; later it partly separates from that system and uses it like a musical instrument. (We have spoken of this when we considered the power of love in the heart.) The image of an instrument goes back to Steiner, for he spoke of the physical body as the 'instrument' of the awareness soul.[20] Later Buehler referred to the physical body in its three aspects as the instrument of the soul.[21] However, the individual organs of the body may also be seen as musical instruments. They are integral parts of the greater instrument that is the physical body, just as individual instruments are part of the greater instrument we call an orchestra.[22] The ego acts as the conductor; in sickness it lowers the baton. The difference compared to musical instruments is that people cannot buy their organs but are themselves very much involved in producing them and have to keep them alive.

The image of a musical instrument clearly demonstrates the consequences of the two directions in which illness develops. As every musician knows, we do not play an instrument really well until we have come free of it. The same applies to our organs as the instruments of soul life.

In the case of organ neurosis (and more profoundly so in organic disease) the players have not been using their instruments well. They have not gained sufficient independent mastery, or the environment has distracted them in their playing. If their instruments are sensitive, bad playing has made them go out of tune. Then the musicians suffer because their instruments are out of tune–or sick people suffer the symptoms arising from the affected organ. They will turn their attention to the instrument and try to retune it. If they cannot manage on their own they will need a physician to help them with their organ neurosis or organic disease.

People suffering from psychoneuroses and psychoses also produce dissonance from their out-of-tune instruments, but they are not aware of the origin of that dissonance. It is possible that they never related properly to their instruments, or they may have given up in resignation because of something or other that has happened. Then the dissonance throws them into uncertainty, agitation, fear or anger. They reproach themselves: I am playing badly, I am a bad musician! Depression develops and–in terms of the illness–they then feel they are bad people. First an organ was out of tune and now this has also happened to the soul. People suffering from psychoneuroses and psychoses also do not gain free and independent mastery of their instruments.

This may be shown in diagrammatic form:

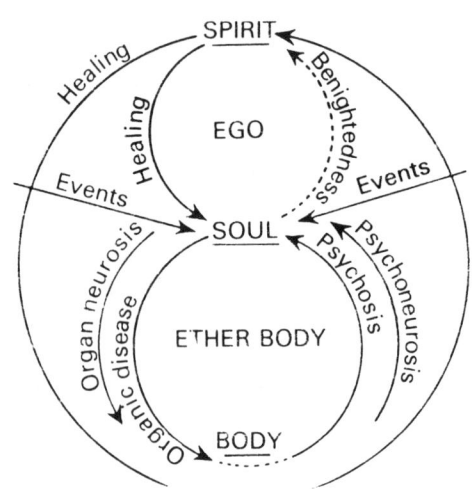

Elucidation. Organ neurosis has its roots in the soul sphere and unfolds in the ether body, and this results in functional disorders in the physical body. Psychoneurosis has its roots in the sphere of the ether body, in functional abnormalities of the body, and unfolds in soul life. This is shown by the shorter arrows on the left and on the right in the lower part of the diagram. The longer arrows in that area represent the more profound processes that lead to organic disease and psychosis. The broken line connecting them at the base of the diagram indicates the way psychosis develops when the organic disease is abating.

The arrows coming in from the sides suggest experiences that on the one side provoke illness right down to the organic level and on the other are not coped with properly because a diseased organ has put the soul out of tune. Organs subject to organic or functional disease no longer vibrate with the soul like the sounding board of an instrument that helps a note to resound fully; instead, the soul has to vibrate with everything that goes on in the diseased organ.

Ether Body and Metamorphosis

Any disorder in body and soul, be it neurotic or psychotic, always means that through the intervention of the ether body a physical organ is also affected. The ether body thus acts as a mediator between the planes of the physical body and the soul, and at the same time also mediates between past and present. The present problem in the soul may be a reiteration of an earlier organic problem, when an organ was 'shaken', or 'concussed' by something the soul experienced.[15] The problem may remain at the functional level or it may progress to organic disease.

At this point, a major law of development is revealed in the rhythmical sequence of the biography. The psychological disorder that develops later on in life sometimes makes its appearance just as many years after the seventh, fourteenth, twenty-first, twenty-eighth or thirty-fifth year as the shock sustained by the organ occurred before that particular nodal point in the seven-year rhythm. When psychological illness develops after the nodal point the ether body of the affected organ 'remembers' and not–as is normal–the ego. Normally the ego recalls something from the ether body of the organ and becomes consciously aware of it. Here, however, the repercussions of the shock originally sustained by the organ come up involuntarily from the ether body into soul life, and the ego submits to the psychological illness that has arisen.

It has been my experience that this law can also be observed to operate when a shock to an organ is later followed by organic disease. The shock, together with the psychological experience, occurred as many years before a nodal point as the onset of organic disease afterwards. (See the example given in the chapter on depression.)

We have seen that incarnation and excarnation in relation to life on earth are the archetypal images behind physical and psychological illness. The question arises if the life of the organism also contains healthy archetypes of the processes that occur in physical and psychological illness. It seems reasonable to say that the recall and perception of memory images by the ego is a process going in the direction from the physical body to the soul sphere. And surely the ego takes the opposite road with every act of will, so that the body serves to bring its will impulses to realization in the world. How does that healthy relationship between body and soul differ from that which exists in illness?

In an earlier chapter we saw that powers of growth become powers of thought when thinking bases itself on recall in the learning process and remembered images are brought back to mind. The powers of growth submit to the essential principle of a soul that is forming images, an ego that is thinking. If they were to introduce the principle of gravity that pertains to the physical body into the life of the soul, thoughts could not take wing and independent thought activity could not arise. That in fact is what happens when the soul congeals and grows 'paralyzed' in endogenous depression, so that even a single train of thought is more or less impossible. A modification of this occurs when the powers of growth make delusions grow in the soul sphere that cannot be grasped by logic. ('Soul growths' like these, with the creative powers of the body coming into full play in the soul, will be considered under the heading of schizophrenia in this book.)

The polar opposite process occurs in an act of will. Now other powers of the ether body mediate the ego's interventions in metabolism, interventions arising in the soul sphere, where the will aspect of the ego has its home. According to Steiner, the ego first of all intervenes in the warmth ether and this establishes a relationship to the warmth organism and the combustion processes in

metabolism that are enhanced when there is an act of will.[23] Again
the ether body brings about a transformation that serves the physical
foundations of will activity, and again the spheres of soul and body
do not mix. We have seen, however, that they do mix when there
is a fever and the body is dissolved through powers of soul. The light
of awareness, which is the life principle in the powers of soul, is not
transformed into warmth of will in that case but condensed to become
the fire of pyrexia. This fire can take firm hold in an inflammation.

The polarity between paralysis and proliferative growth induced
in soul life by the body compares with the polarity between dissolution and combustion induced in the body by the soul. In both instances
the ether body no longer catches and holds, nor does it mediate. The
pathology reveals to us the healing principle of metamorphosis which
is failing to a greater or lesser degree in physical or psychological
illness. With both types of illness metamorphosis is incomplete, and
the inherent principles of one sphere come into play in another.

Thus we see that the inherent principles of soul and body must
remain separate.[7] The only connection between them should be
through transformation or metamorphosis.

The Spirit – Disease and Healing

The diagram on page 140 was based on humans beings having body,
soul and spirit, and this already pointed to the theme of this section.
The process representing physical illness does not arise in the spiritual
principle, not does the process representing psychological illness reach
the spiritual principle. The arrow on the right in the upper half of
the diagram has been drawn as a broken line to indicate this; it also
shows that the spirit may be 'clouded' or 'benighted' by a
psychological condition and may then withdraw from body and soul.
This brings us to the second maxim in the anthroposophical approach
to psychiatry: The spirit cannot be sick, it is always healthy.[14]
(Many years later, Jaspers came to the same realization from another
point of view.[24])

When we consider the issue of 'spirit and psychological illness'
the following image may arise, and it is one that has also proved
therapeutic in many cases of depression. The sun (representing the
spirit) may be obscured by smoke and vapours rising from the earth
(representing the body), but these cannot touch it. When the earth

has ceased to produce things that obscure it–medical treatment should bring this about if we think in terms of the body–the sun will shine again, perhaps even more brightly and strongly than before. Unlike the sun, however, the atmosphere (representing the sphere of the soul) is indeed polluted and poisoned by what comes from the earth (the body). Compared to the light of the sun, which is only prevented from reaching us and taking effect, the substance of the air is changed.

The soul does not fall sick from causes arising within itself, the way the body does, but it also does not remain free from sickness the way the spirit does. In this respect, too, the soul holds a middle position between body and spirit.

Physicians and other helpers see direct corroboration of the above image when a glimmer of egoity appears again in the eyes of the sufferer where before only darkness of spirit and lack of egoity had shown themselves, and when the chaos of the soul lightens before the eye of the beholder.

Healing is part of the inherent nature of illness and of the human being and it is essentially self-healing; it can therefore only arise from the one principle in the human being that is always healthy–the principle of the spirit. This is indicated by the short arrow on the left in the diagram that goes from the spirit to the soul, just as the arrow that forms a circle around the whole arises from the spirit and encompasses the whole human being. Treatment and healing involve the whole human being, just as the illness does, but they originate in the spirit. In the light of the spirit we can perceive that there are three aspects to therapy that relate to body, soul and spirit.

Three Aspects to Therapy

Prior to undertaking treatment the essential question is: Is treatment really justifiable if one is thinking in terms of destiny? If the ego is not merely accepting the illness but essentially wants it, are we justified in giving treatment to take it away? Then, however, a further question arises: Does not therapy belong to the disease the way a key does to the lock? Is not the patient's ego also seeking the therapy? The therapist is also sought out, after all. Patients often feel that now they have found their true physician or therapist, and such feelings seem to indicate that there is a connection.

On the other hand we are considering a therapy that definitely does not 'take away' the patient's disease the way it is done when psychotropic drugs are given in massive doses to suppress a depression or neurosis. Compared to this, a form of therapy that supports the patient and encourages self-healing helps individuals to come to themselves, which after all is also the aim of the disease. (See the section on Treatment at the Physical Level, below, and Therapeutic Aspects in the chapter on schizophrenia.)

Apart from this a new aspect emerges with regard to preventive measures in the sphere of the soul. It is possible to intercept a physical illness (that might later develop into psychological illness) at the soul level. The 'self-education' it is designed to encourage can be achieved by the ego before the illness develops, and the education and training we give our children can pave the way for this. This is the way of preventing both physical and psychological illness.[25] A preventative approach that addresses itself to body, soul and spirit will therefore aim to combine the use of medicines and other measures for prevention at the physical level with prevention at the level of soul and spirit.

Therapy at the Level of the Spirit
Aspects of therapy for the spirit have already been discussed under the heading of developmental disorders and neurotic syndromes. This form of therapy is only possible if psychology and psychiatry are expanded to encompass the spirit. In the same way we shall be unable to do justice to psychological conditions unless the centre and goal of soul development is taken into consideration and allowed to come into play. Treatment limited to bringing the soul's burdens to awareness remains at the level of the soul, penetrating neither to the physical basis nor to the spirit as the source of healing. The spirit lives in us through the ego, and treatment must therefore in the final instance address itself to the healthy ego of the patient, clearing the way to let it intervene in soul and body. This means that therapy at the level of the spirit is in fact ego-therapy, though this should not be taken to mean that the ego itself is to be treated.

Several schools of psychotherapy have aimed in this direction. Jung went in this direction with his principle of therapeutic individuation and Frankl with his logotherapy, which has been discussed earlier. However, the ego can only become a concrete factor with full therapeutic effect if consideration is also given to its true home, the source of its healing powers. That home is the world of the spirit, and any form of therapy at the level of the spirit, every therapist for whom that world has become a reality, goes beyond the soul aspect of the patient to connect with the world of the spirit.

Therapy at the level of the spirit first of all concentrates on efforts to discern the origin of the illness in the biography, for this may also give the illness meaning with regard to the future. In conjunction with this, awareness of the patient's mission in life is strengthened or aroused; this is something that only reaches its culmination in the light of the awareness soul in mid-life, but will also help recovery at an earlier period. This form of therapeutic dialogue always takes its guidance from the principle we discussed in the chapter on the awareness soul: The light of perceptive understanding in the head must combine with warmth of heart if a healing relationship is to develop between therapist and patient. At the same time there must be complete freedom. There must be no attempt at 'conversion', and nothing should be offered that the patient does not–consciously or unconsciously–ask for. (See Heide and Priever.[26])

The meaning of life and its mission are also a social issue, and in many cases dialogue with the patient needs to be complemented through dialogue with the patient's social partners; the patient's 'social body' also needs treatment. Family therapy is often indicated; if it should prove impracticable efforts should be made to talk to at least one member of the family. Group therapy also has a role to play, as patients are able to learn how to make human contact within a group. If a group session is given a specific theme such as 'anxiety' or 'art therapy', there is less risk of getting stuck in personal concerns; anything of a highly personal and indeed intimate nature does not belong in the group.

There are many patients with whom this form of therapy is not possible. In cases of acute or severe chronic psychosis it may have even more negative effects than in the case of juvenile developmental disorders. Mental and spiritual contents that the ego is no longer able to assimilate may give rise to further delusions or hallucinations. In this case, patients must expressly be prevented from taking in such contents by going to lectures or reading books, and also from meditation.

Therapy at the level of the spirit can nevertheless be used to good effect in these cases, though it would not be mediated in words. Here the therapist's own approach to things of the spirit and his or her personal convictions are a therapeutic factor. Subliminally, people suffering from psychological illness are particularly sensitive to what others think of them and how they speak to them. If the therapist is convinced that the patient's ego is not sick in itself but merely benighted and removed to a distance, no words are needed for the patient to feel that this ego is being addressed as a healthy ego.

Therapy at the Physical Level
This form of therapy bases itself on the second expansion that is needed in psychology and psychiatry; expansion in the direction of the body. Therapy at the physical level is the basis for any therapy of psychological disorders; in the treatment of psychoses it is of prime importance. The question arises as to whether we are not abandoning the principle of 'healing out of the spirit' at this point, and whether the arrow encompassing the whole system in the above diagram may still be said to be appropriate.

Treatment at the physical level primarily involves medical treatment based on the anthroposophical approach to medicine. Anthroposophical drugs have been developed on the principle that it is not only the psychological environment that can provide healing impulses; the physical environment provides natural substances as medicaments, though their medicinal actions can only be properly understood if perception is again expanded and goes beyond the presenting illness, beyond the position of the moment, this time far back into the earliest times of human evolution.

Basing oneself also on observable phenomena, it is possible to realize that the cosmos and the natural substances it produces were much more closely related to human beings in the past than they are today. To find themselves, humans had to forgo that close connection with the cosmos. Growing independent of it and its products, they developed organs with powers that increasingly came to serve a life lived in conscious awareness. If, however, illness has caused human beings to come away too far from the cosmos and such of its powers as are still effective, a medicinal substance will restore some of the true original health. Specific substances have a healing relationship to specific organic processes because in earliest times they were connected with those processes. They now offer a certain potential for the ego to begin new development, this time on an organic basis, for through the substance the organ can take the cosmic archetype for its orientation. Principles that were forgone at an earlier stage now return to human beings as medicinal substances.

Associations like these can only be perceived through a science of the spirit, and this alone entitles us to consider medical treatment, too, as therapy out of the spirit; medicinal substances here take the place of the word. The efficacy of medical treatment is objectively enhanced the more the therapist is able to relate to it in spirit, and this is another way in which the spirit is involved in the therapy which in the final instance serves the patient's spirit. Combining therapy with medicinal substances with therapy through the word, therapists may come to intuit the way the word they speak out of the spirit is attuned to the word of creation that is active in the medicament. (The author discusses another, individual aspect of this in Husemann/Wolff, *The Anthroposophical Approach to Medicine* vol.2, chapter on *Treating the Body*.)

Within this general context it should also be mentioned that with serious psychiatric illnesses it is often necessary to prepare and loosen up the physical ground for effective medical treatment by the extensive use of hyperthermal baths, febrile temperatures induced by other means, short courses of insulin therapy, enterocleaners, venesection etc. as well as work and a healthy diet.

Therapy at the Level of the Soul
Art therapy holds a middle position between the other two therapies. It really is psychotherapy in the proper sense, for it addresses the experiencing soul most directly. A special chapter on art therapy will follow the discussion of individual syndromes.

Television, radio and cinema films are the complete opposites of art therapy. Used for entertainment purposes they, too, address the experiencing soul, but achieve the opposite of creative, activating artistic work and active occupation with works of art. People suffering from psychological conditions in particular grow more and more passive the greater the dependence which developes in this area, i.e. they will increasingly seek to fill their inner emptiness with a flood of images and sounds and to suppress the anxiety-inducing experiences of their condition with the diverting and exciting contents provided by the mass media. It is evident that any existing weakness of the ego will be further enhanced. Initially the ego will not have sufficient time to assimilate the impressions gained in front of the small screen, later the ego will not even have the strength to do so. The same, however, will then also apply to artistic impressions gained in therapy, for it will no longer be able to do anything with them.

It is known that radio and television induce social isolation, and in these patients they will enhance existing contact disorders. Quite apart from the fact that the contents are generally unsuitable we must also consider the physical effects of long periods of television-viewing. The subtle organic changes seen in patients with psychological conditions are bound to be particularly affected by the negative effects that television has on the circulation. The hollowed-out sensation that is experienced after watching television for an extended period and can be felt also at the physical level is evidence of the fact that the constituent principles are no longer fully engaged. A situation has arisen that may well contribute to the development or aggravation of a psychological condition. To complement any kind of art therapy, and all other forms of treatment, these patients should as far as possible keep away from the above mass media or be kept away from them.[27]

Chapter Five
Psychiatric Conditions in Relation to Biography

The psychiatric conditions discussed in this chapter are first described in the terminology generally employed in this field and then expanded and deepened. To begin with, the main characteristics of the particular condition are developed on the basis of typical case histories to establish the clinical picture. This is then discussed and the disease is related to the seven-year periods in life. Distinction will have to be made between congenital traits that may also have a hereditary element, and causative factors in the environment. (The common background to both was discussed in the chapter 'Birth of the Ego and Soul Development'.) It will also be found that there is a distinction between the form a disease takes in childhood and the form it takes in youth and adulthood. Within those age groups diseases generally develop during specific seven-year periods, and the time of onset also reveals something of the character of those periods.

The physical bases of psychiatric conditions are discussed briefly. For a more detailed and systematic presentation of these, and of medical treatment and remedial eurythmy, see the author's chapter on psychiatry in Husemann/Wolff, *The Anthroposophical Approach to Medicine* vol. 3.[1]

Obsessive Compulsive Reaction

Clinical Picture
A primary school teacher in his mid-twenties presented himself in the neurologist's surgery. He looked older than his years, was of

slender, asthenic build, and appeared anxious, dejected and sometimes abashed. Asked what had brought him to the surgery, he immediately gave his diagnosis. He was suffering from a compulsion neurosis that he simply could not cope with. He had an obsession that his hands were dirty and had to keep washing them many times throughout the day, always for minutes at a time. Initially breaks between lessons had been adequate for this, but now it would happen that he had to go to the cloakroom to wash his hands even during lessons. Suppression of the compulsion caused difficulties with breathing and made him so restless and anxious that he was hardly able to continue with the lesson. During lessons he took care to touch things as little as possible, avoiding above all the dirty forms the children sat on. His class–a relatively uncomplicated one–was beginning to notice, with obvious consequences during lessons.

A remarkable element in the history was a very strict mother. As far as the patient was able to remember she took care–even punishing him if necessary–that everything about him was 'decent' and he kept clean. He soon got in the habit, on going to bed at night, of placing his clothes in the exact sequence in which he would be putting them on in the morning. Later it was always right leg and right arm first. Sometimes the mental concept 'sock' or 'trousers' would persist in his head, he had to keep thinking of the object or mentally say its name. Even as a child he had been very anxious to 'do everything properly' and had also been nervous of people. He had been rather sensitive when playing with other boys and easily hurt. Later he tended to brood, finding it impossible to let go of feelings of hurt and of injustices in the world. He would often feel sad, yet at the same time was ambitious and always near the top of the class. On the other hand he was almost all the time afraid of school and unable to shake off what had happened at school. For a time he had to take care on his way to school that a certain house, or a street corner, was reached with a definite number of steps taken, otherwise things would happen to him at school.

He had had a dislike for dirt even as a child. He took care to stay as clean as possible and always washed very thoroughly. The physical aspects of puberty seemd dirty to him. He felt the same about jokes made on the subject. Then it happened that he fell in love with a girl–from a distance, of course–more a form of worship, he insisted–but others spoke of this friendship as dirty. This had been passed on to him and had been the biggest shock of his young years, making him feel profoundly insecure and sad. During that time he often dreamed of snakes in a dirty pond, and from then on he withdrew even more from his human environment, looking for and finding solace in nature and in art.

Washing his hands, always a frequent action, then became more and more frequent. Now his doctor had given him a certificate because he was unable to continue teaching. He felt terrible about being a failure in his profession, particularly as he could perfectly well see that the action was senseless. Previously he had not been so clear about this, but he had consulted books on the subject: Even if you washed your hands for hours, they would never be entirely clean, and that was why surgeons put on rubber gloves after washing their hands. To some extent people had to put up with dirt. He knew all this perfectly well–but he still could not rid himself of the obsession that he was dirty and the compulsion to wash.

The Pathological Process
The genesis of the compulsive state of the patient pointed first of all to his mother; through her, compulsion had first taken hold of the child's soul. This developed into the archetypal experience, later to become conscious: There has to be order in the world. Clothes are put on in strict sequence, and the first compulsive ideas arise, with the child unable to escape from them for some time. Later he became hypersensitive and anxious with reference to certain impressions, and particularly the impression of dirtiness. It happens with many of these patients that the lack of order in the world and in the final instance their own life, whose course they are unable to perceive, condense into such an impression. In puberty this was experienced in his own body, taking the form of an image in his dreams of snakes.

It would be wrong, however, to generalize concerning such factors as a strict mother and repressive upbringing. Some patients have grown up in very favourable conditions but nevertheless show a congenital hypersensitivity and anxiety. In conjunction with their hypersensitivity, perceptions made in the environment take as firm a hold as they do in patients with acquired hypersensitivity. Even relatively minor negative elements that cause only minor and short-lived reactions in other children and young people cause them to experience profound shocks, as we have seen in the above history. With all patients suffering from obsessive compulsive reactions consideration should however be given to a possible contributory role played by more marked negative effects from the environment created by civilization. (See the next-but-one section on 'The Illness in Relation to Biography'.)

It is also possible for young people to grow more aware of their sexual development in puberty and suppress this at the conscious level, thus providing a further factor in the urge to get 'clean' by washing. An external action takes on symbolic character, as we have also seen in the above history, on the way to school. An unfounded compulsive fear of school was fought with symbolic actions performed in lieu of resistance to what happened at school when such resistance was not possibble. Patients will sometimes be tormented by the compulsion to attack someone with a knife, and this may express hidden feelings of aggression towards that person. All this, and much else, is however less important that the initial hypersensitivity. These patients are, or have grown, hypersensitive to impressions and this causes the mental images arising from perceptions made in the world or in relation to their own bodies to become obsessional ideas. Kretschmer quite rightly spoke of the 'sensitive personality type' as the basis of obsessive and compulsive states.[2]

Hypersensitivity means excessive awareness in sentient life; hyperexcitability may be the response. We have seen that lack of sensitivity is a withdrawal to the head, with coldness developing in sentient life. Hypersensitivity on the other hand arises as awareness shifts from the head to the sphere of sentient life. The feeling astral body is caught up in excessive awareness of the world and feels exposed to the world even before it comes free in puberty. If this gets worse, the astral body will no longer be able to let go of outside impressions and forget them. Instead of being 'breathed in' in a healthy way, those impressions are more or less held fast in the life of feeling in the middle region of the human being by a hypersensitive astral body that has gone into 'spasm'. Being held in that sphere of semi-awareness, they return and impose themselves again and again on conscious awareness.

The result is that the astral body does not unite with the idea in sympathy but rather tries, again and again, to reject it out of antipathy. The idea feels like a foreign body to the astral body which resists it and tries to keep it at a distance, yet cannot get free of it. Critical detachment which may even go as far as total insight into the condition–as in the above history–is a key sign for the diagnosis of obsessive compulsive disease in adults. It differentiates compulsive ideas from delusions, for patients usually identify with these.

The intolerable tension produced by the experience of a foreign body in the life of feelings creates the characteristic anxiety felt by obsessed patients: environmental anxiety concerning old and new outside impressions. If the anxiety separates from the idea, anxiety states known as anxiety neuroses develop. These, too, are excessive reactions to outer events that happened a longer or shorter time ago; there always is, or was, a particular idea in the background.[4]

When the anxiety of obsessive patients turns into fear of something or other, this kind of phobia may be considered a form of anxiety neurosis (like the cardiac phobia discussed in the chapter on neurocirculatory asthenia). In the history given above, the compulsive process had advanced even further, i.e. into the will sphere. The idea of dirtiness was first of all transplanted from the head to the hypersensitive life of feelings and the compulsive idea combined with fear of touching anything (haptophobia). In the sphere of the will this developed into the compulsive act of washing his hands, the compulsion to wash, so that in spite of the patient's insight, acts of will were subordinate to the compulsive idea. Critical detachment differentiates this type of compulsive act from symbolic acts that are equally compulsive but at most have a certain self-critical attitude as a background element. This marks the borderline to superstition which, like symbolic acts, has an illusory as well as as compulsive element. People performing symbolic acts will for instance desperately maintain the illusion that the act will prove helpful. (Concerning illusions, see the chapter on hysteria.)

The ego is more or less helpless when the astral body goes into this kind of spasm. Normally the ego is relatively free in the head and more or less free in its manipulaation of concepts; with phobias it is carried along by the astral body and goes with the concepts that arise in the life of feelings, and there it is compelled to enter into an association with them that is much too intense and based on personal feeling responses. Another form of compulsive illness that in practice plays less of a role is the polar opposite of the above in so far as it takes place in the conceptual sphere of the head. It may be briefly outlined as follows. The compulsion to ask questions arises, with the patient compelled to move from one concept to another in the form of questions. Compared to the immersion in the inner life of the soul that has been described above, the ego does not enter sufficiently into the thinking process and is therefore unable to put a stop to the sequence of concepts and stay with one particular answer.[4a] A weakness in the will aspect of the ego is common to both forms of the disease, expressing itself in the sphere of thought in the one and that of feeling in the other.

Organic Aspects

With reference to the breathing process in the soul it should be noted that obsessive patients have the emphasis on the antipathetic gesture of exhalation, but the exhalation gets stuck and does not continue on into the feeling of liberation that can be experienced towards the end of the expiratory phase at the soul as well as the physical level. This characterization strongly suggests a connection between obsessive compulsive illness and pulmonary respiration. Physical breathing problems and shortness of breath are frequently found in obsessive patients–as in the history given above–and may combine with precordial pain. A connection may thus be seen with bronchial asthma.

'Asthmatics often show obsessive traits.'[5] Physical and emotional hypersensitivity can be a major factor in bronchial asthma, presenting as an allergy to inhaled airborne materials. Shock may be another triggering factor. After such an experience, the consequences of which extend even to the body, the disease may shift more to the physical level so that asthma develops. If it shifts more to the soul level, obsessive traits develop. In both cases the experience is that a foreign body has entered that cannot be properly exhaled. In the case of asthma, the air passages go into spasm around the inhaled air, so that exhalation cannot be achieved. In the case of obsessive illness, the astral body takes a spastic hold on the compulsive idea in its vain endeavour to get rid of it. In both cases the spasm originates in the astral body which is unable to let go of the air passages and the stream of air in asthma, and unable to let go of the idea in obsessive compulsive states.[6] In the light of these phenomena obsessive neurosis may also be called asthma of the soul.

The nature of the compulsive ideas also suggests a central relationship with the lung. They are the 'hardest' soul elements known to psychiatrists, and the patients are often themselves aware of the unyielding and unchangeable nature of their compulsions and describe them as such. Again it is the lung that is responsible, for its influence combines with the tendencies that ideas or mental images have to develop a life of their own.

The lung is not only the organ of respiration and thus the basis for the sensitive life of feelings; it is also the organ of the earth element, the solid element, and out of this a melancholic temperament may develop–as in the case of the above patient.[7] (In the case of obsessive compulsive states, this temperament takes an extreme pathological form in a tendency to brood and remain caught up in past events.) The air organism of the body is connected with the renal system; the lung is the organ, the gate, for inhaled and exhaled air. As the physical body opens up to the air through the lung, its essential character is in polar opposition to the essential character of the airy element. The lung is very much a formed-out organ–this reveals a relationship to the head–and more than any other organ is inclined to harden, to mineralize, calcify and even produce bony structures. We are thus able to see that the lung is a central organ for the fundamental power of the ether body that consists in taking organic structures into solid form and out again (life ether).[8]

In this context we can also see what Steiner meant when he said that powers like these that gather in the soul may loosen their connection with the physical body and become involved in the development of compulsive ideas. (It is because of this loosening process that we consider obsessive compulsive illness to come under the heading of psychosis rather than neurosis, though it is usually referred to as compulsive neurosis or anxiety neurosis.) Only brief reference can be made to an even more complex aspect, which is that the real function of these powers is to create the head in the next incarnation. This adds a further nuance to the way the feeling life becomes subject to head forces in obsessive patients.

The Illness in Relation to Biography
'Onset is frequently in childhood, and the condition usually develops before the age of 25.'[9] The patient whose history is given above also showed a childhood form of obsessive compulsive illness in the first seven-year period; causative environmental factors certainly always go back to early childhood, and a childhood form of the illness may develop at that stage. In terms of human development this is connected with the way children are largely determined by the head in their perceptions, and in obsessive compulsive states this has become pathological. During the first seven-year period this does not relate so much to hypersensitivity but to the fact that the impressions mediated via the head are of a consistency that makes them foreign bodies in the child's soul life. They cannot be 'incorporated' in the

body and become 'organic impressions', and therefore also do not stimulate creative or constructive processes at the physical level. Such impressions result from being 'broken in' rather than brought up– as in the above case history. Other factors arise through the whole 'unnatural way of life' in modern civilization; this may provide a basis for foreign bodies in the soul and consequently inadequate physical development.[10]

The consequences are seen above all in the lower region of the organism. According to Steiner, obsessive illness in children bases on 'inadequate development in the system of limbs and metabolism' which then also is no longer able to receive favourable or neutral influences. This weakness shows itself clearly at the physical level in the asthenic constitution of adults–as in the above case history. As the production of protein (low in sulphur) is impeded even during the first seven-year period, impressions cannot go down below and be forgotten even at that stage. They continue to radiate back to the head from where they had originated. It is thus not only the nature of the impressions but also the organism itself that causes the problem, for it has been made in such a way–partly due also to unnatural impressions leaving their imprint–that it more or less has to reject the mental images that are wanting to go down. The children then cannot shake off impressions and will be compelled to repeat over and over again such phrases as 'That's a nice clock!'[11]

Differentiated obsessive illness will usually only develop in the second seven-year period in children who are 'often highly intelligent'.[12] It is only now that the astral body, acting out of the lung, makes the differentiated feeling-based contact with the environment that may lead to hypersensitivity and compulsive symbolic acts. Overwhelming feelings of anxiety cause children to attach particular significance to certain ideas; e.g. that a particular number of paces taken means that all will be well at school. Bringing the idea to realization in a symbolic act that may also be in lieu of something else, children try to banish their anxieties.

If hypersensitivity develops as early as the first seven-year period, this is a case of precocious development. Children who are subject to compulsions at that age also tend to look older than they are. The same may apply to adults, as in the case history given above. Quite generally the impression one gets with obsessive compulsive conditions is that trends belonging to later stages are emerging too early. An inclination to be orderly usually shows itself in the fifth seven-year period or later; in these patients it becomes compulsive pedantry. The critical period of the ninth year seems to play a special role in the second seven-year period.[12] If the ego of a child does not fully enter into the (poorly developed) lower region of the organism, the life of feeling will initially preponderate, and this may grow hypersensitive through impulses coming from the head. If the forming of mental images in the head gains marked ascendancy over a lower pole that could not be properly reached, obsessive compulsive illness develops with weakness of will and passivity towards mental images formed with reference to the outside world. The constitution of early childhood has united with the situation that arose in the second seven-year period, and the inability to forget based on metabolism has united with the hypersensitivity in the soul life that is connected with the lung. Impressions are not sufficiently taken up by the physical body and at the same time the astral body is unable to let go of them. It has already been said that Steiner spoke of the 'soreness of soul' of hypersensitive children. He went on to say that if this could not be improved, hysteria would develop as the 'female form of this soreness' (see next chapter). He also spoke of a 'male form' which might also be discussed.[13] Considering the soreness of soul that persists throughout the history of obsessive patients (and of hysterics), it is possible to conclude that Steiner meant obsessive compulsive illness; in his next lecture in the course on remedial education he discussed this, though without reference to soreness of soul.

Obsessive-compulsive illness has more of a male character (though it occurs in both sexes), and this will become more evident when we come to discuss hysteria. Apart from anything else the marked tendency to hardening in the disease process points to the male organism, for there the tendency is more marked than in the female organism and goes hand in hand with letting the head rule the soul. Hardening is in evidence in the nature of compulsive ideas. And now, when adulthood has been reached, critical detachment from those hardened compulsive ideas shows itself. This detachment is still very much in the background with the softer or more diffuse compulsive ideas and symbolic acts of childhood, and it again steps into the background with the anxiety neuroses of adults.

Now and then many of us may note the first stages of compulsive illness in ourselves. It happens, for example, when we are subject to a compulsive tidiness that we ourselves feel to be a compulsion. If one puts a hand into the post-box after putting in a letter, to make sure the letter has really dropped down, this means one is excessively tidy. If one does it a number of times and even then does not feel quite sure, it is time to do something about such an obsessive-compulsive condition.

Therapeutic Aspects

As with all psychiatric conditions, basic therapy should consist in medical treatment and remedial eurythmy that initially addresses the organic level. With patients whose condition is not very severe and who cannot come to the surgery very often, I have seen definite results even with this approach on its own. In more serious cases, and in less severe cases that have become chronic, treatment has to be more intensive, possibly also on an in-patient basis, and this combines the above with treatment addressed to the soul.

Below, some examples of treatment addressing soul and spirit will be given where therapy takes its orientation from the position of the illness in the biography. This is also the method of choice in milder cases.

It has already been shown that people suffering from severe psychological conditions must be met at the point where they actually are. We may start with the inclination to be excessively accurate and tidy and try to transform it into a more acceptable form. Regular and careful observations of the natural world can teach these patients to combine accuracy with animation without feeling under any compulsion or developing antipathies. The pathological trait will then begin to have positive results. (This can only be done by people whose condition is less severe; seriously ill people would initially feel subject to compulsion with such exercises.)

Animation will increase if not just individual observations but whole pictures are recreated from memory. In creating such pictures out of their own resources, patients slowly get out of the habit of passively giving themselves over to outside impressions. The tied and bound imagination of obsessive patients is further stimulated when they start to paint with water colours as part of their art therapy. Hymnic music and suitable speech therapy have a liberating effect also on the breathing. Exercises like these help patients to catch up a little on the art teaching they did not have or that was not using the right approach in their second seven-year period. They learn to move more in harmony with the world.

The will element in the forming of mental images and ideas can be influenced particularly by deliberately producing the opposite of a compulsive idea in one's thoughts. Steiner suggested 'counter ideas' in such situations that should not be normal either but 'go to the other extreme'.[14] The patient whose history was given above should therefore be encouraged to visualize his hands a brilliant white, with not a particle of dirt to be seen, something that never really happens. (He would of course be unable to do this whilst in the state in which he initially presented.)

Patients use their will power even more by making their hands dirty on purpose, e.g. by putting them in dirty water. This is an exercise that can prove highly effective at some later stage in the treatment of phobias; it bases on Frankl's 'paradoxical intention' which consists in desiring or undertaking the very thing one is afraid of. If you have to keep going back to make sure the door is closed, say to yourself: 'If that door is standing open, let it stand open. I don't mind if they take all the furniture.'[15] In my view, it is not the wish that is central, but rather the will impulse of the ego that completely changes the direction of the fearful withdrawal, i.e. brings to realization or lets happen the very thing the soul wants to run away from in its fear. This allows the beleaguered soul to exhale a great breath of relief, and with it comes the surprised realization: Believe it or not, I have actually made my hands deliberately dirty, when normally I cannot stop washing them. If one is then able to have a laugh about it, real progress will have been made.

In the case of agoraphobia, the fear of crossing an open space, the idea of the emptiness of that space is so powerful that the will is paralyzed and the individual is unable to cross that space. If patients first try to see themselves crossing just a small part of that space the legs will be able to follow that line of thought and the first steps are taken. Stage two consists in again visualizing the next bit, and so on.

If compulsive questions arise, the ego may try to intercept by asking: Is there any real point to this? Inner compulsions to repeat a word or a tune are sometimes overcome by saying the word out aloud several times or singing the tune.

Generally speaking, patients will get furthest in getting their souls out of the prison their obsessive-compulsive illness has created if they try–if possible right from the beginning–to rise above personal concerns and enter into the non-personal realm of the spirit. They need to catch up on the active expansion of interest that engages the whole human being and has somehow been missed in an education that was intellectual and addressed itself onesidedly to the head.

Individuals who overcome obsessive-compulsive illness grow more active in their response to outside impressions. Nervous fixation on those impressions may be transformed into accuracy and due care. The disease will then be seen to be meaningful for soul development, for this can now come free of the body in a healthy way.

Hysteria

Clinical Picture
A neurologist was called to the bedside of an office worker who was in her early twenties. She told him that her right leg had been paralyzed for weeks and that she had no sensation in it. She had woken with pain in her right leg one morning; the pain had passed but the paralysis had developed instead. Asked what had happened before that, she spoke of friction in the office. Some of her colleagues had complained that she got on their nerves because she talked so much. No one had considered that things might get on her nerves, too. Then the head of the department had asked her to come and see him and she had to admit that she had been afraid of the interview. There was a chance that he would dismiss her. However, before the appointed day arrived she had fallen ill. Of course they would say that she was not really ill but only pretending, so that she need not go and see the boss. But that was not true, her leg was really paralyzed, and an examination would show that it was.

Examination revealed that something was definitely wrong with her leg. The right lower leg was slightly swollen compared to the left and felt cooler than the left leg. Sensation of touch was reduced: The patient thought she could only vaguely feel something when the skin was touched from above the knee and down the leg. A striking feature was that the young woman, who was of ample proportions particularly in the region of the trunk, kept looking at the leg herself and plaintively telling the physician that it was like a dead leg that was no longer part of her.

By now she had grown very excited and it was necessary to reassure her that she certainly was not pretending. Taking the history, it was evident that her childhood had been disturbed by many moves. She had found is particularly difficult to cope with a change of school at the age of eight. At the new school the others had ragged her a great deal and this had depressed her. She had then withdrawn and told herself 'stories', a kind of fairy tales in which she was the princess. On one occasion she told a friend about this, also mentioning that her family tree included members of the nobility so that she really had blue blood in her veins. This had caused a great deal of tittle-tattle, though to be honest she had enjoyed this. Later she had had 'super girl friends' and they had been as thick as thieves. On the other hand there had always been disappointments, and then they had really hated one another. Boys and later men had certainly interested her, she had to admit this, but she had never 'had relations' with any of them. She had always sought their company, and altogether needed company now that she was grown up. Her parents had been very good to her; they had 'really spoiled' her. She kept in close touch with them, but there was nothing they could have done for her in the office situation.

At the physical level she reported digestive problems; sensation of fullness after meals and constipation. She was often chilly, though she also perspired a great deal and suffered from cold hands and feet. Her periods started late and were infrequent. She had wet her bed as a child. If she got a sore throat and a cough she would always run a temperature.

The Pathological Process

Hysterical symptoms developed even in childhood. Bedwetting was the first sign. According to Steiner, hysterical bedwetting and other symptoms of childhood hysteria are due to 'the astral body leaking out' which might also show itself in increased (cold) sweats. This was in anticipation of part of what happened on death, when the astral body separated from the body. The 'soreness of soul' of a child that we have already spoken of combines with this, and the image is one not only of the sensitivity of but also of bleeding from the wound. Such children connect too much with the outside world in the leaking of their astral bodies and develop too much awareness in their sentient life. Their hypersensitivity relates not so much to sensory impressions but rather to the soul element that is active in them and to the natural elements: Gravity, the fluid element, the air and heat in the environment.[13] Hysterical children are therefore sensitive to much of what goes on in the souls around them and remains unspoken. They are sensitive to climatic and temperature changes and soon feel too hot or too cold, a problem that still existed for the above patient.

The question as to how the astral body came to leak out took us to the parents who had spoiled the patient as a child. This child did not withdraw into herself in response to excessive warmth in the family nest but enjoyed it; the foundations were laid that would later cause the leaking out. However, the fact that leaking and hypersensitivity also develop when the family situation is different–favourable or unfavourable–suggests that there is a congenital factor.

In the above case history as in the case of the obsessive patient, hypersensitivity and shocks marked the beginning; yet the soul life shows quite a different reaction. Instead of withdrawal into extreme broodiness that was soon to be followed by compulsive ideas, this patient withdrew into wishful thinking and then also presented this to the outside world. Her withdrawal may therefore be considered a 'forward flight'.

The patient enjoyed the fuss she created around her, and this is a feature one often sees in other hysterics. The environment must always be involved, so that they can use it as a mirror in which to experience themselves. It is easy to see that–in contrast to the fear of the environment felt by obsessive patients–this may develop into dependence on the environment with the social problems this creates.

Compared to the obsessive-compulsive soul life that occurs as a transitional stage in many sensitive subjects, we have here a soul life of illusion in its early stage which need not be pathological. In an obsessive-compulsive soul life the inner soul is taken captive by the reality of the world; soul life in illusion changes inner reality to suit its own desires. (This also contributes an illusory element to symbolic acts and superstitions.) Obsessive people are governed by antipathy, people given up to illusion live out their sympathies. This may go so far that someone who has grown hysterical even enjoys pain.

Organic Aspects
According to Steiner, the lung is also involved when illusions are formed.[16] With compulsive ideas the solid element is dominant in the soul; with illusions one would think more of the airy element. The astral body gives itself up to this element at the level of the lung, i.e. in contact with the outside world; it 'builds castles in the air', as the saying goes, most appropriately. The solid element shows itself in the definite contours and consistent features of those castles in

the air. (The person concerned is always the princess or the prince.) This may even take the form of fixed ideas, as sympathy for a dominant, favourite idea threatens to rigidify. ('The only way to peace for the world is through a world government.')

We have seen that obsessive soul life relates to asthma; soul life given up to illusion tends to relate more to inflammatory disease of the respiratory tract. It has been shown that dedication to the world may go to the extreme of a 'fever of the soul'. Devotion to illusory ideas that relate to the world may 'inflame' the soul. At the physical level one may then see inflammatory conditions of the respiratory tract. This connection has been little explored so far, even less so than asthma. A soul life given to illusion is a highly characteristic feature of tuberculosis, which is a chronic inflammation of the lung. Considering the essential nature of the human being, one sees excessive antipathy shown by the astral body as it holds the process of expiration in spasms at one extreme, and its excessive sympathy coming to expression in inflammation at the other, so that the latter may be regarded as a pathologically deep 'inhalation' of the astral body.[17] (See also the chapter on *Physical Illness Originating in the Soul*.) The process may be enhanced so that a temperature develops–as in the above case history.

Soul life given up to illusion–and one that is governed by obsession–will however only progress to definite psychological illness when a wrong attitude of soul that initially was of short duration becomes a determining principle to which both soul and ego are subject. In obsessive-compulsive illness the soul is one-sidedly determined by the outside world; hysteria develops when the soul tries to shape its relationship to the outside world one-sidedly from within. It then aims to bring its illusory wishful ideas to realization, whereas before it had merely given itself up to them inwardly.

That is how the clinical picture presented by the above patient arose. The wish not to have to go for an interview of which she was afraid connected with the memory picture of a paralyzed leg. It is indeed part of normal soul life that our wishes connect with such images. For a brief moment we heave an inward sigh, thinking that if only we were paralyzed, we would not have to go for that interview. A hysterical subject is no longer able to put such a wishful notion aside, so that it is 'immediately taken hold of by the surging swell of feelings', 'by the surging swell of feelings in the organism'.[18] The illusory wishful idea is brought to realization by an astral body that wants and desires paralysis and so a form of paralysis does indeed develop.

In principle hysterical paralysis is a process of organ neurosis: Soul elements acting via the ether body enter into and disrupt the function of an organ, i.e. its motive processes and the flow in its water organism. Experiencing this functional disorder, which also affects the perceptive faculties of the skin, the patient enters into it to such an extent that she is no longer able to move the leg and loses sensation in it. Compared to organic paralysis due to inflammation of nerves or muscles, the solid substance of nerves and muscles is not affected; instead, circulation is partly blocked in the leg and its nerve tissues, and the leg grows colder. Another feature is that the paralysis always takes the form imagined by the patient, particularly where loss of sensation is concerned. The area affected by this is not the area that would be affected if there was organic damage to the nerves.

With regard to soul life we realize that the patient's perceptive and image-forming soul and ego faculties were fixed on the leg, so that it felt like a dead object to her. She may have been exaggerating, but this is indeed what happens with any paralysis. The perceptions of the ego in particular are focussed on the paralyzed limb as if this were an outside object to which the perceptive faculties are directed; it is a feeling known to everyone who suffers from paralysis. The pain felt in the initial stage arose specifically as the idea-forming astral body entered into the limb, though this does not go as far as it does in a case of organic paralysis.[19]

Compared to organic paralysis and non-hysterical organ neurosis, this is a case where the wishful desire of the astral body has given rise to the condition; flight into an illness that the sufferer herself created dissolved the fear. Such a flight may also cause a rush of movement rather than paralysis; in that case the astral body is acting from below, i.e. from the kidney system. This may also lead to (non-epileptic) convulsions or seizures. Whichever this may be, hysterical subjects are genuinely ill. The difference to simulation is that in that case the aim is only to deceive the world, whilst hysterical subjects also deceive themselves.

Some hysterics set whole dramas in scene around them in their desire to bring their wishes and illusory ideas to realization. In the final instance hysterics try to shape the world around them so that it corresponds to the desires of the astral body. Their demonstrativeness and indeed highly theatrical behaviour–as in the above case history–must partly be seen as an appeal to others to join in the production. Another part of the way they work themselves up may be due to the fact that hysterics always feel insecure, that they have a vague notion that their behaviour is artificial rather than genuine and this makes them put it on even more.

Those are also the symptoms of the kind of 'hysterical reaction' many people show in situations that are particularly shocking or exciting, though this does not make them hysterics in the pathological sense. (They should nevertheless do something about it.) Here it shows itself that the human constitution is such that the first beginnings of hysteria are in the astral body, which dominates the reaction in situations like these. On the other hand symptoms like these may have their root in a hysterical nature that is always present and may give rise to pathological hysterical reactions. Hysterical individuals need not always escape into physical illness; they may also withdraw into twilight states. The dreamlike state that is normal only for the life of feelings then occupies the whole of soul life.

The question arises as to how the astral body comes to be so dominant and where the source lies of creative powers that make the lramatic productions created by hysterics seem almost like works of genius at times. Again the answer will be found if we consider the illness in relation to biography.

The Illness in Relation to Biography
Obsessive-compulsive illness was shown to have its origins in the first seven-year period. Childhood hysteria develops essentially in the second seven-year period; one hardly ever sees hysterical symptoms prior to three or four years of age.[20] Bedwetting does go further back, but it does not have the same demonstrative hysterical character. The majority of hysterics are initially young people at or around puberty.[21] This is the time when the astral body develops and transition occurs to the female form of soreness, i.e. adult hysteria, which then shows itself 'particularly in the third decade of life'.[21]

Distinction must be made between childhood hysteria in the second seven-year period, which may be due to environmental factors in the first seven-year period, and the hysteria that develops in puberty and adult life.[22] The image of 'the astral body leaking out' suggests that in this case the astral body is still active in the etheric life that is active in the watery element in the seven-year period. People who develop hysteria during or after puberty have souls that effervesce, flicker and flame so that one experiences the gales and thunderclouds of an openly exposed astral body. The hypersensitivity of hysterical children–Steiner gave a description of this in his *Curative Education*–is directed more to the elements than to the people around them. As puberty approaches, however, 'hysterical disorders are noted in behaviour and in the body'.[20] One sometimes has the impression that puberty occurs precociously at the soul level in these children and that the astral body starts to act independently out of the reproductive organs before the time is ripe.

In the soul, this premature birth of the astral body leads to precociousness in the life of feeling which then shows certain traits belonging to puberty. These children behave like teenagers and therefore also look older than they are. Premature erotic sexuality may develop in the soul sphere. Steiner emphasized that sexuality is not part of childhood hysteria; yet it may show itself as a soul factor in the second seven-year period if puberty is preempted. Sometimes this may in fact be induced in a child by other people.

In the body, premature birth of the astral body may combine with delay in physical maturation, particularly in the genital sphere.[21] In typical cases of hysteria–e.g. in the above history–the body conformation may lack form (dysplasia) and there may be a tendency to swell up and exude fluids; hysterical children on the other hand show a more graceful slenderness, so that they tend to look delicate. This difference again shows the effect that puberty has on hysterical subjects. Prior to puberty the astral body entered into the lower region of the organism; during puberty it rises from it. The body becomes more rounded as the astral body enters, but in hysterical children leaking of the astral body has caused the process of building the body to be weakened from the middle region. In hysterical adults on the other hand the middle lacks form because astral body and ego do not come in sufficiently to mould it; the astral body continues to leak, but then from the lower region of the body; the physical habit follows that trend. This is the clinical picture that Steiner called 'hysteria' in the wider sense, a constitutional process where 'metabolic functions grow too independent' and this extends even to 'sexual symptoms' and digestive disorders.[23]

Apart from physical conditions such as inflammations, such a hysterical constition may also give rise to hysteria as a psychological condition that has been preconditioned by the leaking that occurred in childhood. This takes us back to the archetypal phenomenon of excarnation. The reproductive organs are retarded in their function, and sometimes also in their growth, whilst sexual powers show themselves in the life of the soul. This corresponds to the way Hippocrates, the early Greek physician, saw hysteria: The womb (*hystera* in Greek) roamed or pushed its way around in the body. This refers to the dynamics, of course, i.e. to powers of the astral body. Due to their leaking out in childhood, to errors in upbringing, shocks etc. these have not fully incarnated as the reproductive organs came to physical development. Instead of a healthy birth process one sees the liberation of powers that should have remained active in the organic sphere. According to Steiner, many of the traits that show themselves in the soul life of hysterics are due to 'metabolic processes in the reproductive organs'.[24] 'Suppressed sexual longings' may then come to expression in 'profoundly mystical drawings and paintings.'[25] Powers of sexuality that do not come to fulfilment in the body by contributing to the creation of a new body become active and indeed creative in the sphere to which they have risen.

This extends far beyond the sexual sphere, however. We have seen that the powers of imagination that are liberated at puberty relate to powers of growth in the lower region of the organism and particularly the reproductive organs. These creative organic powers that initially have nothing to do with sexual desire move up into the life of the soul at puberty. They partly use certain wishful images to combine with the powers of sexual desire in the astral body and help to make us procreative in the reproductive sense, and partly become creative by being transformed into powers of artistic imagination.

This transformation does not take place, or remains partial, in hysterical subjects. The powers of creative imagination retain the characteristics that belong to the procreative faculty of sexual powers and serve to bring the egotistical desires of the astral body to realization in the environment, i.e. in the patient's 'social body'. These powers, which in the physical sphere primarily show a certain 'genius', are responsible for the brilliance of the dramatic productions created by hysterics. Those productions arise therefore from abnormal creative imagination, i.e. again an incomplete metamorphosis of physical powers into powers of soul.

In biographical terms we can see that these patients get stuck in this particular situation. In a hysterical character, 'residues from the stage of early puberty' persist throughout life.[21] Hysterical children show teenage traits during the second seven-year period, when they should still be children. Hysterical adults more or less remain teeny-boppers for the rest of their lives. The term has been chosen to indicate that females are more commonly affected, and this seems reasonable if one considers the emphasis on the life of feeling in the female character. Men suffering from hysteria–and these do of course exist–show particular feminine traits.

Compared to obsessive-compulsive illness, which originates in the head (and therefore comes under the heading of neurasthenic conditions), hysteria relates to the abdominal region and particularly the sphere of the reproductive organs. Their powers rise and enter into the illusory soul life of the lung, which leads to illness. Steiner suggested that in hysterical children the surfaces of the organs were too porous so that the astral body leaked out. This fits in with the principle according to which the lung is constructed, for not only does it show a tendency to mineralize but its internal surfaces are designed to let the airy element pass through. We may imagine that because a hysterical subject has remained more or less stationary at the stage of lung development, some of the porousness of the lung also develops in other organs. This applies particularly to the urogenital system, from where the astral body with its wants and desires pours out into the world via the lung. This may be experienced as a pathological enhancement of the sanguine temperament which has a certain excitability and instability.

Due to the dominance of the astral body, hysterical subjects are not sufficiently centred and have not sufficient egoity. This points to the 'heart' as the ego organ, which takes second place to the lung in hysterics. There is a weakness of heart forces in the soul that shows itself in the emotional instability of these patients who may be hot or cold in both body and soul but never show steady warmth. Hysterical people do not relate to others from the heart; they flow out into the environment via the lung. Their 'love' is not for the other person as he or she really is, but for the dream image of the other person that is their own creation. Disappointment may cause such 'love' to turn into hatred.

It is also possible for obsessive to change to hysterical behaviour and vice versa, or there may be a mixture of the two. Generally, however, the emphasis is on one pole or the other.

With regard to soul development, the difference may be summarized as follows. People suffering from hysteria or illusional states retain something of the youthfulness of their puberty in later life, both inwardly and in outer terms. In people suffering from obsessive and compulsory illness an ageing process becomes fixed; this originates from the head and its effects go back to the first seven-year period and sometimes also show themselves in the outer appearance.

Therapeutic Aspects
Medical treatment and remedial eurythmy– only in very serious cases on an inpatient basis–again provide the basis. In mild cases medical treatment may sometimes be sufficient on its own. In more severe cases medical treatment and remedial eurythmy will shorten the period when treatment addressing soul and spirit will be required, and above all make the results more lasting. Dedicated psychological treatment can also achieve lasting results at the psychological level, but there is the risk of a conversion reaction. This represents a symptom shift from the soul sphere to the untreated physical region; sometimes circulatory and metabolic disorders may develop only at a later stage, when the connection with the 'cured' psychological condition is not perceived.

Exercise therapy acts in an area that lies between body and soul. It was used with the patient whose history is given above. Through soul and ego it addresses the 'ground-in' disorder in the ether body. Medical treatment and remedial eurythmy release the astral body from its fixation on the leg and prepare the lower region of the organism for proper incarnation; graduated movement and walking exercises teach the ego and astral body to intervene in a healthy way and let go again in the process of movement. Such essentially physical exercises also play a role in other physical disorders of hysterical subjects.

At the soul level we take our starting point from the hyperactivity of the astral body and the misuse of powers of imagination. If hysterical subjects are allowed to play a role in a theatre production, or at least give a recitation, they are provided with a forum where they can justifiably bring their powers of imagination into play and let the astral body be active. Entering into their stage roles they no longer feel the need to play-act in life and can now to some extent catch up on the transformation of a subjective imagination that served to fulfil personal desires, into artistic imagination. Such creative activity also gets the perceptive and form-giving faculties of the ego more involved. Once again a pathological disposition can be turned to positive account.

Life has to be given to the imagination of obsessive patients, but form is needed in hysterics. This gives the orientation for all artistic exercises, where the direction should always be from the outside to the inside. Modelling will have an effect on the rather porous and sometimes over-abounding constitution and help patients to cope better with soul powers that tend to flow out. The transition from major to minor in music therapy encourages a growing inwardness of body and soul, so that patients are better able to come to themselves. Speech exercises address particularly the ego weakness (see the chapter on Art Therapy.)

When treating hysterical disorders it is most important not to limit oneself to the phenomena the disease presents but expand in the direction of both body and spirit. Expansion in the direction of the body by considering the organic aspect has already been discussed; it also provides the basis for a more objective assessment of hysterical subjects. Expansion in the direction of the spirit is sometimes actually longed for by the patients. The hidden longing for the spirit that we have spoken of may be present and cause additional unrest in the soul sphere. It may however also lose itself in mystical aspirations and pathological visionary experiences. This is where it is particularly important to heed Steiner's advice and base oneself on training the faculty of thought. In more serious cases, however, the therapy described above will be needed first to put patients in a position where they can learn to think. If one tries to push patients in this direction before that point is reached, or if they try and force themselves, one may find that they are tackling the theory of knowledge in a hysterical way, waxing lyrical about it and enjoying it without having really grasped it and made it their own.

When the biography is discussed, it is important to direct attention to the disappointments or disillusionments these patients have suffered and set in train. One idea is to start by discussing the meaning of the verb to 'dis-illusion' which literally means getting rid of illusion. If one looks at it from this point of view one may actually come to feel grateful for such disappointments or disillusionments. Realizing how they deceived themselves in their illusions, individuals may find it easier not to create illusions for others.

Our own attitude to hysterical subjects reveals the degree to which we ourselves have been caught up with them at the soul level. No other psychological condition gives such an immediate experience of the astral body of the other person, and none is as infectious as hysteria. It will of course depend on the situation in our own astral bodies whether we grow hysterical ourselves. Sometimes we merely get caught up in the excessive sympathy process of hysteria. We will then identify with our hysterical patients, finding everything they produce so interesting that they will justifiably consider this an invitation to launch new productions. Psychoanalytical interpretation sometimes goes to such extremes that one gets an impression of personal pleasures being satisfied in scientific guise. It is self-evident that such an attitude will not help our patients, however great our sympathy may be.

Another danger that exists at the subjective level is that we react to patients with antipathy, giving in to the annoyance they are certainly able to cause in more than full measure at times. So we give them a set-down, only to find that they are worse. Here we perceive the difference between appropriate and inappropriate shocks administered as part of treatment. If a shock is given out of anger, i.e. out of an involuntary reaction at the soul level, it will have negative effects. If the same shock is applied out of the ego and in a desire to help, i.e. out of love, it may prove medicinal.

It is of course totally reprehensible to use the word 'hysteria' as a pejorative. Hysteria is always a pathological state or process. Hysterical patients, or people showing hysterical reactions, are essentially asking for help; a proper understanding of the disease process helps us to give them that help or encourage them to help themselves.

Individuals who overcome the pathological process of hysteria may have transformed their hypersensitivity and outflowing qualities into sensitiveness and imagination in projecting themselves into the minds of others. Receptiveness for art and artistic activity now becomes possible and is at a deeper level. Having recovered, they learn to use their lungs as instruments of the soul.[26] Their illness will then have had meaning, also with regard to the future.

Aspects of anorexia nervosa
Seen against the background of biography, anorexia nervosa initially shows a definite relationship to hysteria. There is a constitutional relationship in so far as anorexia nervosa aims at the opposite of the

super-abounding adult hysterical type. The aim to is achieve the graceful 'ethereal' characteristics of the hysterical child. It develops in puberty, sometimes as a reaction to the puppy fat that many anorexics show at the age of 12-14 and to the obesity that is common in the families of these young people. In conjunction with this anorexics resist the physical changes that come with puberty and sexuality itself. They want to return to the second or the first seven-year period but problems exist even there, problems that suggest incomplete incarnation. Earlier bonding to the parents may persist at a subliminal level. There tends to be a lack of harmony in their families and a domineering father or mother who inhibits complete incarnation.

The condition is increasingly common and affects the female sex even more markedly than hysteria does. Females enter less deeply into the lower organism; they find it more difficult to relate to this part of the organism and therefore are more apt to hark back to the life of the rhythmical system. Up to a certain point this may be a transitional stage in female puberty. If it grows pathological, a temporary longing felt in the soul becomes addictive in an astral body that shies away from earthly maturity. The subconscious, illusional wish to return to childhood also influences the physical body. The connection with hysteria becomes apparent, and this sometimes also comes to expression in hysterical traits. On the whole, however, a compulsive and autistic element predominates and combines with the disease process becoming frozen.

Periods tend to stop even before emaciation develops. Sometimes they started rather early, as a sign that the astral body was separating from the lower organism prematurely. The astral body then separates so far from the body that after incomplete maturation the function of the reproductive organs is also impaired and periods stop. This combines with persistent constipation. Emaciation is achieved by fasting, laxative abuse and forced vomiting after meals. Body weight may be reduced to 25 kg, which is a life-threatening situation.

Anorexic girls usually refuse to eat at mealtimes and then consume large quantities of food in secret. Revolted at their own greediness and afraid of gaining weight they then induce vomiting and use laxatives. Again a greedy desire for food develops. A tragedy is enacted by the astral body which is torn between inclination and disinclination and gets caught up in a vicious circle.

The dichotomy characteristic of puberty is enhanced (in some cases leading to schizophrenia). On the one hand these patients often show above-average intelligence, on the other we experience the emotional unrest rising from below that may become a hyper-social outgoing from the middle region which, however, is merely superficial. The breathing of the soul in the middle region also appears to show a split. Thus one will note this out-flowing and a tendency to illusion, and on the other hand also compulsive tidiness. Hypersensitivity may form the background–as in hysteria and obsessive-compulsive states. Above all however there is fear of incarnating fully in the world, of being tightly imprisoned in the body, and this may drive these desperately ill girls to their death.

Treatment has to be on an inpatient basis for more severe cases and essentially consists of medical treatment and remedial eurythmy to help incarnation. At the psychological level, some of the things suggested for the treatment of obsessive-compulsive illness will apply, particularly with regard to art therapy. Therapeutic dialogue with an element of exercise to it will help to achieve maturation (Bockemuehl[27]). The most important aspect is that the ego's will to incarnate is addressed to counteract the astral body's desire to escape from the body. A good way of starting is to present a number of job profiles, so that the will can awaken through feeling. With reference to the job of 'wife and mother', contact with children is recommended. Sick children in particular will help maternal powers to ripen in anorexic girls and these will then guide them to maturity for life on earth.

Epilepsy

Clinical Picture

The patient was a man in his early thirties, athletic build, bloated face, indolent eyes and slow, thick speech, again and again getting caught up in a single word. His wife did most of the talking. Giving the history, she said that her husband had suffered from epilepsy from the age of nine. A paternal uncle had also suffered from attacks. Her husband had been quiet and rather slow as a child, and this had increased until the attacks started for no apparent reason when he was nine. He was soon given antiepileptic drugs. Initially the attacks had got better, but her husband had got slower and slower and increasingly more sluggish in his thinking. He had initially been on phenobarbitone. When this was replaced with other drugs he did not slow down so much. On the other hand his attacks became more frequent, so that the dosage of the new drugs had to be greatly increased. Now her husband had few attacks but was so slow and forgetful that he could no longer follow his occupation as a joiner.

Attacks often developed at night, when he was asleep, though they could also come by day. When an attack started her husband would cry out, and if the attack was during the day he would fall to the floor with his whole body in convulsion. As he lay on the floor there would be regular contractions affecting the whole body. After about a minute the body would go slack and he would fall into a deep sleep. As he said himself, he knew nothing about what happened during an attack. In recent years he sometimes saw flashes of light or heard bells ring before an attack. He felt that he was floating and had a feeling of profound happiness. After that he would know nothing more.

The wife said that on the days preceding an attack her husband would get slower and slower and more or more sluggish in his thinking; he would also be in a bad mood and rather irritable. If there had been a longer interval since the last attack–currently he had one every three months–he would feel exhausted at first but then he would be much better than in the weeks that went before. He would also be less slow then and able to think better. This was not the same with more frequent attacks. Sometimes she would be thinking that he was about to have an attack but he would merely get in a rage; he would really go on the rampage, and after that he was a bit better again. There were also times, however, when he was not quite all there, and he would be going about and behave in an utterly senseless way. Sometimes this condition would only last a few moments, and he would be staring in front of him and not respond. There had been times when he had got crazy notions in his head, saying that he was being followed in the street. With this he would be sad and miserable for days and did not bother much with his work, though he had always needed help with this anyway. He had only been able to keep his joinery workshop going until recently because he had always put a great deal of effort into it and had had constant help.

The patient nodded in agreement with everything his wife said. He thought that everything was due to the many tablets he had to take now. Weren't there some other medicines that would get rid of the disease? He very much wanted to go on working. As he said those words he raised his eyes, which until then had been lowered, and looked shyly and yet with dull hope into the physician's eyes. It was like a boy turning to an adult for help in a situation that is far from clear to him.

The Pathological Process

Convulsive seizures are the outstanding features of epilepsy, and people often mistake them for the disease when in fact they are merely one of the symptoms. Convulsions may also develop in a case of poisoning, for instance, which has nothing to do with epilepsy. Infantile spasms usually point to a disposition for epilepsy and in some cases there is also a family history, as in the above example. The disease itself becomes manifest in the intervals between attacks. The archetypal phenomenon, which is very marked in characteristic cases, is a sluggish stagnation in both physical and mental mobility with a tendency to hold on to single words or thoughts. This tendency tends to be enhanced in the days preceding an attack and will sometimes improve afterwards. Frenzied rages may have a similarly beneficial effect. Stagnation will also be found in metabolism, circulation and respiration, among other things in weight gain due to water retention and oliguria, and in retention of carbon dioxide and a number of substances that are no longer eliminated. This type of metabolism is normal during sleep, though to a less marked degree, and goes to extremes in the hibernation of animals.[28]

The tendency to stagnate may be further enhanced immediately preceding an attack. Respiration slows down and even comes to a standstill, the slow heart beat becomes even slower, the blood pressure drops further, and muscles lose tone. The impression is of sleepiness turning into an abnormal sleep state that goes in the direction of entering into the final sleep. Not only the ego and the soul separate partly from this organism in a way similar to sleep; it is evident from the changes in water metabolism and in the circulation that the ether body also wants to separate, which otherwise happens when we die. The astral activity that continues to maintain the life of the body during sleep also threatens to succumb, as is evident from the drop in blood pressure and respiratory standstill.

Convulsive seizures bring a dramatic change. Lack of tone turns into maximum tension (tonic stage), twitching of muscles (clonic stage), a marked rise in blood pressure and acceleration of heart beats. Convulsions are followed by profoundly deepened respiration, and the blueness of the face that started to develop prior to the attack and got much more marked during it gives way to a healthy pink. Eliminatory functions get going again after an attack and laboratory values for substances retained in the organism are 'closer to normal' after an attack. The explosive event of the attack has partly overcome the stagnation.

What has happened? If we follow such a process as human beings and not only scientists with expert knowledge of the brain and metabolism, if we enter into the sufferings of the individual and his struggles, we cannot be satisfied with the explanation that control mechanisms in the brain failed and that the brain of an epileptic subject is unable to 'maintain a middle position where thresholds and function are concerned'.[28] Control or steering failure–but where does this leave the steersman? If we see the brain to have automatic steering, like some kind of computer, then there is indeed no room left for the suffering soul and the struggling ego of the human being.

Steiner presented a different, more human view that also relates to personal experience. Every morning, when we wake up, ego and astral body enter again into the body. However, if they are to make connection with the outside world and in this way achieve the full awareness of daytime consciousness, they must penetrate the body once they have entered. If there are obstacles in the body they cannot achieve that penetration, or only partly achieve it. They then push their way out in a convulsive seizure.'[29]

It is good to try and see where the beginnings of every disease lie in ourselves, so that we may gain a better understanding of our patients. We have already done so in the case of obsessive-compulsive and hysterical reactions. The beginnings of epilepsy lie at the point of waking from sleep, a moment when we may be aware of a certain muscular tension. Twitching may occur particularly if one has had a 'falling' dream, 'falling' into the physical body. We open our eyes, though we are not yet fully awake, and for a moment do not recognize our surroundings. Apart from the very beginning of a convulsive seizure we thus also know something of the twilight state.

Let us now return to the patient; he can never come completely awake and according to Steiner has got stuck in the waking-up process. The morning twilight in his soul may grow denser and become a twilight state where he meanders about as if in a dream. Such states also occur in greatly abbreviated form as absences that take only seconds. The waking-up process may slide back into sleep–via the twilight state–and in extreme cases this threatens to become the final sleep. We perceive the resignation of ego and astral body as they abandon an incomplete incarnation stage and proceed towards excarnation. During the short aura that sometimes precedes an attack, the loosening of those higher principles induces the beginnings of a psychosis with hallucinations; auras may also appear on their own (in the above case history the persecution complex occasionally shown by the patient). The floating sensation that according to the patient went with a great feeling of happiness is the most direct expression of the loosening process.

We are now in a better position to follow the events of an attack. The patient's own astral body strikes like lightning so that he falls to the ground and it takes hold of the body in maximum tension. Carrying the ego with it, the astral body has initiated a new, forced incarnation. When this tonic phase is followed by a clonic phase with twitching, we may get the impression that the astral body is shaking and knocking against the inner obstacles in the stagnant physical body. This is an implosion where the whole of soul life, forced in upon itself and wholly unconscious in the darkness of the body, exhausts itself in a tremendous effort, after which it withdraws into sleep again, though this is now a healthier sleep. After this, the sufferer may come to a healthier awakening in a physical body where the astral body has partly overcome the stagnation.

Seen like this, an epileptic attack is no longer merely a symptom but an attempt at self-healing, albeit a 'convulsive' one that does not get far beyond the initial stages. The resolution of stagnation through the attack is only partial and of limited duration. At the same time the astral body may grow so exhausted after repeated attacks that less and less power is available for conscious soul life. Last but not least the brain as the instrument of conscious soul life may suffer so much damage that increasing slowness, lack of concentration and forgetfulness arise not just as the outcome of epilepsy but also due to brain damage. (This is known as mental deterioration with an organic base as distinct from character changes relating to epilepsy. More recently many authors have cast doubt on the existence of those character changes or denied them altogether. I myself share the opinion of those who maintain that they exist.[29a])

Even with these limitations it should not be forgotten that epileptic attacks do not represent the disease. Conventional medicine also speaks of polar opposites with regard to convulsions. Selbach for instance speaks of play and counterplay 'between reduced consciousness and the development of motor unrest'. He refers to a fact that is utilized in therapy, that an artificially induced convulsive seizure will interrupt a twilight state.[28] In all forms of motor activity, in extreme cases raging frenzies and convulsive seizures, the astral body tries to incarnate more fully in the body and overcome the organic resignation that then supervenes again in a new twilight state or psychosis.

Organic Aspects

Steiner's presentation makes it possible to understand the metabolic and cerebral changes in the context of the whole human being. He also traced the building of obstacles to individual organs. According to him, the obstacle itself consists in subtle infiltration with matter, or it is created in the 'etheric elements' of the organs. This primary densification is then reflected in metabolic stagnation. In contradistinction to the porousness of organs and leakage of the astral body that one gets with hysteria, the astral body, again taking the leading role in epilepsy, is blocked, together with the ego, 'beneath the surface of the organs'; the image is one of the organs themselves, and specifically their internal and external surfaces having grown too dense. Depending on which organ is primarily affected, epilepsy takes a particular form in both body and soul life.

If disorder in the water organism predominates, one will first of all think of the liver. Watery congestion, sluggish slowness and dullness, and also a tendency to depressive moods and explosive rages, pointed in this direction in the case of the above patient. The principal elements in the change of character seen in epileptics therefore arise through metabolic disorder in the liver. This is a pathological exaggeration of the phlegmatic temperament. As already stated, the astral body radiates into the lower organism through the kidneys. The abnormalities in internal respiration in the air organism and increased emotionality are connected with the kidneys, as are the forcible intervention of the astral body in convulsive seizures and its resignation in absences, twilight states and psychosis. The kidney system comes to the fore particularly when there is increased emotionality and the above-mentioned disorders of consciousness; it is always involved if there are convulsions (see the chapter on Autonomic Disorders). Particularly emotional patients are more lively than the liver-based epileptics who tend to be sleepy; in their case the change of character is less obvious. Nevertheless, they, too, show a certain stagnation with difficulties in perception and thinking.

In adults, the liver system and the kidney system provide the basis for any kind of epileptic phenomenon. If there are additional obstacles in the lung, abnormal pulmonary respiration and compulsive symptoms may be noted and also an element of heaviness in the body and in soul life. Obstacles in the life of the heart lead to abnormalities in the warmth organism, depression, guilt feelings, and frenzied rages. The reproductive organs may also play a role. Underdevelopment in this region may act as an obstacle and contribute to the epileptic process if the astral body struggles against the obstacle rather than withdrawing from it in a hysterical process.

The brain is always involved in epilepsy. It gives form not only to the body but to every movement in the body and therefore also to the chaotic movement impulses of an astral body that wants to push its way through. The configuration of a convulsion, and this may be triggered by irritation in the brain, arises as the astral body (with the ego) comes in 'from above' in a process which is then held in spasm. The character change we have spoken of also arises partly through the role played by the brain.

On the other hand the actual obstacle may be in the brain. Scar tissue that has formed after brain damage, prevents penetration by the astral body and ego so that full daytime consciousness cannot be achieved. The brain damage that remains then causes 'residual epilepsy'. In this, the brain plays the leading role; in 'genuine epilepsy' it has a less prominent part. In the latter case no abnormality is found in the brain and the emphasis is on abnormalities in the kidney or liver system. (In this respect the diagnosis of 'genuine' epilepsy may well continue to have meaning.) Residual epilepsy is assumed to be partly due to metabolic changes. One and the same scar in exactly the same part of the brain will cause epilepsy in some but not in others. In patients who develop epilepsy one must consider the contributory role of an epileptic constitution that is characterized by metabolic stagnation.[30]

Hallucinations relating to different sensory spheres suggest particular sites of malfunction in the brain in which attacks originate. (In the case of the above patient the indications are that such additional brain damage was sustained at a later point in time.) The location of such sites may be indicated and the diagnosis of epilepsy confirmed among other things by an analysis of the electrical currents produced in the brain (EEG = electroencephalogram).

The Illness in Relation to Biography

Like hysteria, epilepsy in adults relates to the second seven-year period, and that was the time when the above patient developed the disease. Care must be taken to differentiate this from childhood epilepsy which tends to have its onset in the first seven-year period and may later develop into adult epilepsy. Children live from the head, so that residual epilepsy is typical in their case. Brain damage suffered during pregnancy or in the birth process, brain haemorrhages due to head injuries and the sequelae of meningitis have their consequences even in the first seven-year period. They may however also create a disposition for an epilepsy that only develops at a later stage.

Brain symptoms play a greater role in childhood than in adult epilepsy. The soul symptoms of physical organs are less to the fore. With the astral body not yet born and not even preparing for birth, the effects of organic abnormalities on soul life are more diffuse and less differentiated. Organic abnormalities and their effects on soul life only need to be more seriously taken into account in children who show precocious development.

Epileptic children thus do not show the psychological symptoms one sees in adults but rather the vertigo, nausea and abnormal states of consciousness that Steiner spoke of in his lectures on curative education. Hysterical children show hypersensitivity to the outer elements; epileptic children are held fast in their bodies and do not relate sufficiently to the elements. Inability to penetrate the lung as the organ of the solid element causes vertigo. (Vertigo results from failure to come to terms with the forces of gravity and achieve a balance between gravity and buoyancy.) If stagnation occurs in the liver as the central organ of the watery element, the nausea characteristic of the liver develops. Stagnation in the kidney system leads to the abovementioned abnormal states of consciousness in the astral body. The latter may lead to 'petit mal epilepsy', with frequent absences where the astral body withdraws from the kidney systems of these children, who are emotional in their responses. (The three special forms of epilepsy seen in children and young people, with petit mal epilepsy occurring mainly in the second seven-year period, are described by Holtzapfel.[31])

'The majority of epilepsies' are manifest by the time puberty is reached.[32] Adult epilepsy characteristically has its onset at the beginning of the second seven-year period. Here the situation arising in the ninth year again plays the leading role; it also tends to present itself again if onset of the disease is at a later stage. We see now that the ninth year is a point on which everything turns in psychological conditions that may arise later or continue on from childhood. On the one hand development may not continue on into the body, so that hysteria or obsessive-compulsive states develop. On the other hand it may continue on into the body but get stuck in the body, resulting in epilepsy. This has its organic basis in the liver and kidney system in the lower region of the organism which is the goal of incarnation in the second seven-year period.

Some of these children will be found to grow progressively quieter and duller in around the ninth year, and this may get worse, as in the case of the above patient, and lead to the slowing-down and the abnormal states of consciousness of epilepsy. Apart from getting stuck in the body, i.e. in terms of space, we also see that these patients get stuck in their biographies. We have seen how the road to puberty leads out of the darkness of metabolism and on the one hand into the light of conscious awareness as the head is lifted out further, and on the other hand into the will-governed movement of the limbs, taking a route via the muscles that ends in the bones. Epileptics more or less get stuck along this road. On the one hand they do not achieve full conscious awareness, on the other hand they are caught up in the sluggish stagnation of their muscles and their movements lack firmness, or 'bone'.

That is also the way in which the athletic constitution develops. This has the emphasis on muscular development in conjunction with slowness and indeed sluggishness of all physical and psychological motility. This type is most characteristic as the constitutional basis for epilepsy (unless one has the unformed 'dysplastic' constitution). In the type of epilepsy that shows the marks of the liver being involved, the 'tenacity' of the athletic temperament becomes sluggishness.[33] Convulsive attacks, and particularly the twitching of muscles, are due to the astral body's attempts to resolve the situation.

We can see a hint of the epileptic process in normal life when we consider the way young people move in prepuberty–now indolent, now exaggerated and convulsive. The relationship between epilepsy and the second seven-year period becomes even more apparent in the 'stretch' syndrome. Children at this age level tend to stretch and loll about more than adults; it is a tendency we all have in the mornings, before we are quite awake, or when we grow tired. This syndrome may show itself in exaggerated form in epileptics, taking the place of an attack. Metabolic studies have shown that this is another attempt to cope with inner stagnation and inability to breathe.[28] In adults this is merely a temporary problem, but for the above reasons it occupies children to a much greater extent in their second seven-year period. In epileptics, the attempt to waken into their limbs by having a good stretch becomes one of the symptoms of the disease.

The psychological situation of the ninth year tends to persist more or less throughout life in epileptics. The rages acting as discharges in which a child fights against the lethargy the body threatens to impose become 'attacks of frenzied rage' in epilepsy. The dull sadness felt in the ninth year increases to become the depression of epileptic subjects; this seems to be covered by a layer of dullness, i.e. the 'blanket' of the physical body. The inclination to roam that develops with puberty becomes a pathological desire to keep moving in epileptics.

Severely ill epileptics do not achieve full puberty. Underdevelopment often reveals itself well in advance in form of a particular infantility. The wishy-washy face that one sometimes also sees in children after their ninth year indicates lack of differentiation at soul level, and signifies that the astral body is partly remaining in the maternal cocoon and in the etheric sphere. Later on one then sees the 'structure of a grown-up child' in epileptic subjects (Tellenbach).

Courage is a power in the soul that children need if they are to enter fully into their bodies, and it is important to strengthen that courage in the years preceding puberty. Epileptics are struggling even more to enter upon the earth through their bodies and to take hold of the world through them. In every attack they shake and knock against the obstacles in their bodies. Sinking into relaxation and clouding of consciousness, these courageous prisoners are gathering strength for a renewal of the struggle.

If epileptics learn to cope with their illness, particularly also inwardly, the courage and unconditional nature of their struggle at the physical level may be transformed into courage and resolution in their endeavours at the level of mind and spirit. Withdrawal from the body may become a healthy liberation from the body. The illness can be seen to have meaning in terms of further development.

Therapeutic Aspects
Medical treatment is even more important with epilepsy than with
the conditions we have discussed so far; it can achieve considerable
improvement or a cure. Medical treatment addresses both the obstacle
in the organ and the epileptic process itself and leads to the resump-
tion of the waking-up process that had been blocked. Remedial
eurythmy also acts in those two directions. To give an example that
demonstrates the method, the 'E' gesture (pronounced like the first
part of the ai in pain) from the ego to encourage deeper incarnation
and at the same time initiates penetration of the body. The astral
body is blocked in its progress through the body in the waking-up
process, and crossing the hands and feet in the 'E' gesture encourages
continuation and completion of the process. There have been cases
where intense work with that exercise prevented an imminent attack.
This was achieved by relieving the astral body of a difficult task and
thus making the attack unnecessary, which is the aim of a rational
epilepsy therapy which does not involve the suppression of attacks.
    It would be wrong to ignore attacks, however, for they are attempts
at self-healing. The above case history showed that relatively frequent
attacks absorb too much of the soul's energies and damage the brain.
They thus become a second disease. It was noted that the patient
tended to hold on to certain thoughts or words; his second disease
showed a similar tendency to hold on to its 'abnormal rhythm'.[34]
One has the impression that attacks develop because of this, and not
from metabolic necessity. It is therefore important to treat the second
disease as well.
    If there are frequent attacks, however, we cannot wait until the
therapy based on anthroposophical medicine begins to take effect.
Organ therapy in particular has to act right down into the constitution and therefore tends to take a long time; during this time the
brain may suffer irreparable damage from attacks. In such cases,
therefore, where the harm is always greater than the benefit, conventional antiepileptic drugs have to be given to reduce the attacks
before anything else is done. The modern drugs are much more effective than phenobarbitone, which was widely used in the past. They
serve to control all kinds of seizures, suppressing them from the brain.
The greatest danger with these drugs is that they also suppress the
self-healing impulse that forms the basis of any attack, and enhance

the above-mentioned character changes with their typical features of slowing-down and clouding of consciousness. They may also cause cerebral abnormalities.

The conventional school, too, believes that in some cases physicians have only two alternatives–'mental illness or epilepsy (Landolt, 1963), character changes or attacks.'[35] The dosage will therefore be kept as low as possible (though adequate), and this is possible if conventional drug treatment is combined with the therapy outlined above. It will then be found that the damage such drugs can cause is outweighed by the benefit. When attacks are less frequent, on the other hand, we have more time to give real treatment and can do without conventional drugs to begin with–subject to the agreement of the patient and his or her family.

Art therapy should consist in modelling exercises that strongly stimulate form principles to take hold of the stagnant organism. Painting helps to overcome the sluggishness and tendency to go into spasm. Music leads out of the dull moods epileptics develop. Speech exercises have the greatest effect on the ego and on conscious awareness. Wishy-washy speech is given more definite shape, and the rhythms of speech loosen up the immovable rhythms of epileptic subjects.

Therapy for mind and spirit may take direction from the fact that these patients have more or less got stuck in their second seven-year period. In severe cases they need authority wielded with love to help them progress from the dullness of soul life caught up in the body to the discernment and sentience of the third seven-year period. Again is is particularly valuable to encourage the formulation of questions. If the condition is less serious patients may be able to follow this road on their own. It is most important for both patient and family to understand that seizures are not attacks mounted by the body but represent attempts made by the patient's own soul, which is active in the body, to overcome the physical illness. Patients and their families should also know, however, that everything is being done to relieve them of those forceful attempts.[36]

Schizophrenia

Clinical Picture
A student in his mid-twenties came to the surgery saying that he was on the verge of a nervous breakdown. He was of slender, asthenic

build and showed a nervous restlessness. When he grew less restless his movements had a certain stiffness. Mentally he seemed distraught and sometimes profoundly disturbed. His speech was rushed and somewhat monotonous, and again and again there would be twitching of the face and especially the forehead, like distant lightning flashes. His eyes were unsteady and would turn away if one attempted to look into them.

Asked as to what caused the disturbance he said with some hesitation that something very strange was happening to him. He was being pursued wherever he went. He was being watched at the university and in the street, people gave knowing smiles and made comments. When he was on his own he heard the voices of his persecutors, both known and unknown; they talked about him, abused him and gave him orders. Sometimes he would also receive orders in other ways, for the thoughts of people who came towards him in the street would suddenly be in his head and he would have to move the way those thoughts told him. His movements would be getting very stiff then, he had to move in fits and starts and pay attention to his movements. If he followed their orders those people sometimes showed 'satisfaction in their faces'.

Asked why all those people took such an interest in him and what they wanted with him, he said he was not sure. They had probably all heard of the life he led; he had sometimes heard statements made on television that referred to him. The whole thing had started after his divorce. Two years previously he had married a fellow student. They had had common interests relating to their studies and also physically. Then, however, it turned out that they had different life styles. She had been very sociable, whereas he was not at all sociable. She had always wanted to go out, whilst he did not. So they had separated months ago. He supposed that she had been telling some of their friends about their marriage–that was only logical. But that everybody should gradually come to know about him, surely that was going too far.

The patient had been quite cool and indeed almost indifferent when speaking about his marriage and its consequences. However, when he was asked if he had loved his wife the dam burst; he was shaking with emotion as he said that she had been everything to him; he had always wanted, indeed needed, to have her around, he needed her tenderness. This had proved too much for her. And now things were as they had been before; he was on his own again, and spent hours thinking about it; he had to get clear in his mind. But things were not the way they had been before, he felt completely changed, 'like another person'. The disappointment had completely finished him. That was probably what they wanted to find out from him with their experiments. But they knew; he had said so several times, 'I am finished!' He was in a state of great agitation as he produced those words.

A point of interest in the history was that he had an uncle who 'suffered from persecution mania'. Unlike himself, his uncle had imagined all kinds of things. His family home had not been at all bad. His parents had had little time for him, but he had not missed their attention, nor was it true that he later had difficulties in making contact because of this, as psychologists liked to maintain. His uncle, for instance, had had a very happy family life but nevertheless had grown solitary. He himself had not looked for contact even as a child. He had done well at school and been top of the class for many years. He did not make friends, however, because they could not relate to his ideal of friendship, and so he had only known disappointments in this respect. When he felt misunderstood he immediately withdrew into himself. He also had to admit that he often got too agitated, particularly if there was a difference of opinion, and that he had then become aggressive. He found it painful that he could not really express what he felt. The others would say he was too abstract. His studies (German literature and history of art) were no problem, too easy in fact, he had got bored with them. In his school days he had sometimes felt deeply when reading or hearing a poem or a play, but as his studies progressed such feelings came less and less often. Art and literature now more or less left him cold or made him laugh.

Asked if that was all he had to tell, he gave a hestitant denial. His story indicated that he really lived three lives: His life as a student, his life as 'a guinea pig', and a third life. The patient now struck quite a different note; his expression became embarrassed and deferential and then froze in fanaticism. Apart from the voices he had mentioned he sometimes heard one that was different, a voice of thunder that would say: 'Get up! I shall be coming in the clouds!' Sometimes he would see the one who said this. At first he would see only his eyes and these were as if made of light. Then he himself would appear, as Michelangelo had painted him, the judge of the world walking in the clouds. At first he had thought it was a memory image, for the painting had made a great impression on him in the past. But it became more and more real and took form, and there were also flashes of lightning. Christ, the true Christ, was saying that the world would perish and that he would judge it. This was imminent. The end of the world would come in the ninth hour, for that was also the hour when Christ had died on the cross. Whenever he looked at a clock or watch the figure nine had a special lustre. This had become more and more marked recently, so that the figure nine now shone like a star.

Asked if he was afraid of the end of the world he said 'No' with conviction. Quite coolly again: He needed no help in this affair. He just wanted something for his nerves. He could not continue with his studies the way things were.

The Pathological Process

A number of schisms were immediately apparent in the patient's soul life. One even experienced a schism in the way he spoke of himself: He was facing himself in doing so. He had been apart from his human environment even in childhood. It seems that, as in other children with such problems, an inherited disposition combined with parental neglect, though the patient denied this, considering himself definitely above it. The psychiatrist's diagnosis was 'autism', i.e. withdrawl into himself. A further schism was apparent in the inner life of the soul, where a cool intellectual approach with a tendency to abstract thinking contrasted with hot emotionality and increased excitability.

The sentient life also showed schisms. The patient was hypersensitive inside but presented a cool and abstract exterior. This hypersensitivity made him avoid eye contact. We note hypersensitivity in obsessive-compulsive and hysterical reactions, and lack of sensitivity and dullness in epilepsy. Here both poles are present, sometimes with one overlaying the other. The sentient life therefore appears to be inwardly torn in two.

Schism is the archetypal phenomenon in this kind of personality, and the condition is therefore called 'schizoid'. According to Kretschmer, a schizoid personality may form the disposition to schizophrenic psychosis.[33] Many schizophrenic patients show one or the other of these character traits (particularly autism) in the history.

When schizophrenia becomes manifest–the above case history presents the acute picture–we note first of all an increase in schizoid phenomena. The agitation increases, and at the same time intellectual coolness, and a tendency to abstract brooding, are enhanced. If the situation grows more acute the excitement may turn into bouts of frenzied rage, with patients completely beside themselves and a danger to themselves and others (the above patient had almost reached this point). Cool, unmoved observation and abstraction may in turn terminate in phases of stupor, where these patients, having grown insensitive, would lie in their beds without showing any emotional or will reaction and without stirring a muscle for weeks and indeed months unless one did something about it. They are usually perfectly conscious when in this phase and notice everything that goes on around them, having their own (pathological) ideas about it. In frenzied rages, patients are beside themselves; in a stupor they are almost entirely outside themselves, and present only in the perceptive and thought functions of the head. This partial excarnation goes hand in hand with depressed physical functions in the middle and lower organism, i.e. respiration, blood circulation and metabolism.

Such states are beyond empathy and comprehension; they seem to be more elemental and largely beyond psychological understanding. The stupor phase may be compared to the heavy immobility of the sea and the atmosphere before a storm. To use the language of the elements: The principle belonging to the solid element has taken hold of water and air. Then the absolute opposite happens and the patient goes into a frenzied rage; it is like a great storm where all the sharp contrasts that had shown themselves before are fused into one.

This fusion, which may also develop without frenzied rages, points to a further archetypal phenomenon of schizophrenic psychosis–chaos arising in soul life. In the sphere of emotionality and of the will, from where movement impulses arise, patients experience the thoughts of other people; these act down from the head and cause the movements to stiffen or to go in fits and starts. The principle of immobility belonging to the head may however also cause stiffening and obstruction at the lower pole when there is no connection with the environment. Conversely, emotionality has entered into thinking and sensory perception. The above patient's persecution mania had his thinking and sensory perceptions whipped up by excitement and the emotion of fear. The lightning-like flashes in the region of the forehead may be considered an outward sign of this. As the process of dissolution advances, thinking and speech become disjointed, so that associations and ideas, or fragments of them, are presented in a sequence that shows no logical cohesion, sometimes simply as verbal sequences. Words or syllables may be fused and neologisms, or new words, produced. At other times–a sign of rigidity coming in again–words (or movements) are monotonously reiterated over and over again (stereotypy). (See also the chapter on Hoelderlin.)

To sum up: The opposite extremes of soul life come into effect in a heightened way and interpenetrate.

If we ask ourselves what happens in the middle region, the life of feeling and sentience which has been 'passed over', the answer is that 'loss of the middle' takes on extreme forms in this disease. It can be experienced that the ego simply is no longer able to act as a mediator in this elemental process. The result is that it has become difficult if not impossible to achieve a balance between above and below, and the connection to social life that normally arises from the middle is partly or completely inhibited. In the above history this came to expression in the persecution mania, and again this will be easier to understand if we look for the beginnings of delusional states in ourselves.

We have all of us been in a position where people said unkind things behind our backs. Temporary insecurity may occasionally make us think that comments made by people who are less closely connected with us refer to ourselves when in fact this is not possible. How does this get put straight? Our healthy sense of reality tells us that it simply cannot be true that people who hardly know us can talk about us like that; it cannot possibly interest them, etc. The healthy sense of reality is also a social sense that has made us enter into the situation of those other people. Schizophrenics, whose sentient middle sphere was poorly developed even in childhood, have completely lost that sense of reality through their psychosis.

After and during the chaos that develops in soul life, secondary schisms may develop. The above patient lived two lives simultaneously and to him there was no connection between them. He came to the surgery to get help for his nerves so that he could continue with his studies, though at the same time he was firmly convinced that the end of the world was imminent. Schizophrenics will sometimes act in a way that is the direct opposite of what they are thinking. Finally we perceive an ego disorder in so far as these patients not only feel that others control their egos but also experience themselves inside different individuals who may be living or dead. 'Disorder of the person' is diagnosed,[37] or a multiple personality, and these and other ego disorders are considered hallmarks of schizophrenic illness.

Yet there can be no such thing as a splitting of the individual personality, for this is indivisible. The picture of the sun's reflection being 'split up' by the waves on the surface of a lake makes it clear that only ego activity is subject to schism in a soul life that is splitting and loosening up; the ego itself is not. This is the split between thinking, feeling and will activity that we have spoken of; Steiner also referred to this when he discussed 'split personality' in another context.[38] The higher ego, the sun, is left untouched by the process, but its radiation, the lower or earthly ego at the centre of the soul, is more or less lost. When we experience that someone like this is beside himself, we are essentially aware that his ego (the lower ego) has departed from soul life and that only the ego powers that are able to relate to the split are active, powers that can identify with the perceptions of others or the mental images they form. The soul life has remained personal and we get the impression that it is split and in chaos. The ego on the other hand seems literally 'de-ranged', shifted out of its life in the soul. (See the discussion of so-called 'mental' or 'psychological' illnesses in the chapter entitled *Connections between Physical and Psychological Conditions*.)

Considering the above phenomena of schizophrenic psychosis as a whole, one gains the direct impression that it not only involves an aggravation of schizoid personality traits but also shows that something new has come in. A process has begun that partly emphasizes and partly changes soul structure, dissolves it, and then as the disease progresses lets it come to rest again. The soul life of these patients is like a volcano that splits open and ejects a stream of lava that engulfs everything in its path, after which the chaos of the lava, broken up as it is split up further, slowly congeals.

The schizophrenic process may develop in bouts like this, or it may be gradual. In some cases there is just one episode, after which soul life returns completely to its original state. In others the condition progresses until a kind of end state is reached where separation from the world and inner schism have become a permanent and usually more stable state of schizophrenic dementia. These patients are always different from other people, freakish and chronically deranged. Yet even in that state nobody knows if one day the separated ego, the loosened soul, will not regain contact with the world and put an end to the so-called end state. This prospect alone is enough to help us remain patient and give care and help to these patients. If nothing else, such help will bear fruit in a future life on earth.

Essentially schizophrenic psychoses fall into three categories. If changes in perception and logical thinking dominate the picture the diagnosis is paranoid schizophrenia (with delusions and hallucinations). If the changes are more in the life of feeling, with silliness of response and action and shallow emotional reactions, the diagnosis is 'hebephrenia'; in this, the symptoms of puberty tend to be exaggerated. 'Catatonia' is diagnosed when schizophrenic changes affect above all the life of will and of movement and there are phases of tension and relaxation, frenzied rage and stupor.

Organic Aspects
The new element which arises in the schizophrenic process cannot be understood on the basis of soul life, and some patients are actually aware of this. Something alien has irrupted into soul life with dynamics that do not originate in the soul. Could it be that these storms in the soul are connected with the patient's innate air element, which has taken hold of soul life? Perhaps the image of a thunder storm is particularly apt because the frenzied rage in the soul is really and truly an elemental process. In fact, internal respiration has been found to be deficient in some patients who had a tendency to go into frenzied rages–a form of 'internal asphyxiation' that is at a more subtle level than in epilepsy.[39] Could it be that the air hunger is then felt at the level of the soul, and that the forces of the astral body, congested because they are unable to penetrate the air organism properly, have become active in the sphere of the soul and discharge in great storms?

The pathological 'contents' of schizophrenic soul life also point to the 'living body'. What happened when the persecution mania started in the above case history? At the beginning, and this is common in delusional states, there was a real experience. The above patient's divorce led to an understandable fear in the mind of this hypersensitive schizoid individual that his former wife might be talking to mutual friends about things that had happened in their marriage. A sequence of mental images relating to those things developed and was touched on each time the patient saw those friends; every one of these perceptions would be gone over again and lead to further feelings of shame. Up to this point we have been able to follow the process. Then, however, the patient came to feel that all the people he met in the street had heard of his marriage. To begin with he continued to perceive real happenings: Some people would be smiling to themselves for some reason, others might just happen to look satisfied or annoyed, others again would be saying something to their partners–and the patient related all these things to himself.

How did this come about? Initially it was due not to the perceptions but rather the complex of ideas connected with the shame that he experienced. That complex had since grown. The initial fear that even more people might have heard of his marital problems and divorce had given rise to anxiety-inducing notions of the kind that normal people may also have. For the schizophrenic, such notions become reality, i.e. they turn into delusional concepts that come to occupy more and more of the individual's inner life. Many things that the patient perceived and that were indifferent in themselves, were now interpreted in the light of those delusions. He could not help doing this, for the perceptions always touched on the complex that dominated his soul life and were then assimilated in the light of it. Mental images that normally serve as organs of comprehension had combined to form organs for mania and delusion. (See the chapter on *Metamorphosis of Powers of Growth into Powers of Thought.*[39a]

Later the actual perceptions also changed. The patient not only misinterpreted what he heard but misheard things in the light of his persecution mania. Finally he heard voices even when there was no one about. It is evident that all this arose out of himself. His own ideas of shame and anxiety came to expression in the 'perceptions' that came to him from outside, i.e. the organs for the forming of mental images in his soul had undergone pathological changes and were involved in creating those perceptions. Patients will sometimes be aware of the transition from inner to outer life. They will say that initially the images had been so lively that they could be heard as inner voices, and then the voices had come from outside.

Further insight into the new structures that develop in soul life can be gained from the following considerations. First of all the question arises as to how sequences of mental images grow in the souls of these patients. Powers of growth are not inherent in the soul and must therefore come from some other source. They are part of the 'unnatural' processes that are not 'natural' to the soul and therefore are also felt to be unnatural. As they are powers of growth they can only originate in the ether body. They are, however, powers of growth that no longer submit to the laws that pertain to the sphere of thought and no longer serve the thinking ego. As previously suggested, they introduce some of their own growth dynamics into soul life, causing delusions to grow and multiply. The result is something we might call a malignant tumour in the soul, a soul cancer with the tendency to fill the life of the soul with its growth and change it just as a cancer changes the physical body. The non-proliferative delusions seen in the majority of non-schizophrenic delusional states may be compared to benign tumours.

The question as to why the ether body's powers of growth have changed their sphere of activity thus leads us to the physical body. 'If we hallucinate this means no more and no less than that the body is sending elements up into conscious awareness that should really be used in the digestion and for growth ...'[40] According to Steiner, delusions are also connected with physical organs,[41] so that a similar process may be envisaged. 'Powers of digestion' are involved in the dissolution of soul life, and powers of growth make delusions and hallucinations grow.

The voices are sometimes as clear as real voices heard in the physical environment, and this is because they are produced by forces that really should be active in the physical body. The same applies to the graphic vision of Christ described in the above case history and other visual hallucinations. Hallucinatory voices are experienced as coming towards the patient because they arise in the sensory sphere of the soul where all contents are experienced as coming from outside. The senses are deceived because a process coming from inside appears to be coming from the opposite direction.

At the same time the whole of the schizophrenic's soul life expands and extends into the surroundings as it loosens. Elements that only lived inside before now also live outside. This is a process in which the inner soul content is partly turned inside out, with the effect that these patients have so much 'about them' that one experiences their anxiety more in the atmosphere around them than inside them.

Another consequence is that the patients themselves are more intensely involved in what goes on around them and proceed to make it part of their delusions. Compared to the loosening of the soul seen in hysterical subjects, the ether body, too, is loosened in schizophrenia, so that hypersensitivity becomes an elemental sharing in experiences. Schizophrenics who are inclined to depression may identify with the withering and dying aspect of nature in autumn to such an extent that they feel responsible for this. Conversely, the withering and dying process in nature may be experienced in spring, and this may develop into an experience of the end of the world that is quite common among schizophrenics. What happens is that the soul and ego expanding into the world experience their own excarnation from earthly life as the end of the world, and again there may be a number of perceptions that are assimilated in the wrong way and reinforce the image. The patient whose history is given above, for example, interpreted the figure nine on a clock or watch to the effect that the world would come to an end at 9 o'clock. We can understand that this would make him feel completely changed. (Concerning the reality of such experiences, see the chapter entitled *Schizophrenia and the Supernatural World*.)

The next question to arise is how the higher principles come to excarnate. This points to the metabolic disorders inherent in schizophrenic psychosis, which can only be mentioned briefly in the present context. Certain symptoms indicate that there are subtle defects in the protein structure of organs; Steiner suggested this in 1921. Having become 'defective', the organs are no longer able to contain the organic powers of the higher principles, and this applies particularly to the creative powers of the ether body. The (unconscious) 'image' of the organ (which determines its growth) can no longer come to expression in 'moulding' the organ and imposes itself on the soul life, where it causes hallucinations.[43] Some of the art work done by psychotic patients gives immediate insights into the creative powers relating to organs; in other cases these creative powers combine with other powers that have come away from the organic sphere to produce the new structures or neoplasms in soul life that we have been discussing.

In some cases, scientists have now demonstrated the excessive degradation of organic protein that leads to the subtle 'deformation' mentioned above. It was also found that in catatonic patients eliminatory functions were depressed and there was retention of protein cleavage products that may be considered to have toxic properties. (Attempts to treat these patients with kidney machines proved partly successful.) In the majority of cases, however, these and other abnormalities are not at a level where quantitative assessment is possible. (It has been possible to trace them to some extent by the method of blood crystallization.) At this point the pathological process on the whole shifts back to the psychological sphere where the tendencies for cleavage and dissolution of schizophrenic metabolism and the subtle chemical dynamics of metabolites with their toxic properties take effect. A schizoid disposition meets them half-way, and a soul life that is already inclined to cleavage or schism opens up to powers of cleavage and dissolution that arise from the patients' own bodies. The close connection between protein metabolism and kidneys, and the eliminatory abnormalities, disorders of the air organism, and the emotional soul life of schizophrenics strongly suggest the kidney system as the organic basis for the loosening of the astral body. Other organic systems are then drawn into the process by the kidneys. Steiner considered abnormalities in lung metabolism to be responsible for certain delusions.[44] If the kidney-based excitement develops into frenzied rages, the heart has become involved and its inner warmth turned into a consuming fire.[16] The liver always plays a part if there is a depressive element. (See the chapter on *Depression*.)

The Illness in Relation to Biography
Schizophrenic psychosis usually becomes manifest between puberty and the 25th (in women the 30th) year. In terms of age, too, schizophrenia is therefore characteristically a psychosis of youth, and this was indicated by the older name given to the disease, 'dementia praecox', which means early or precocious dementia. (Steiner still used this term.) As we have noted, however, only some patients develop defective states and these may be mild or severe and may also revert to normal, so that the term 'schizophrenia', created by Bleuler, is definitely the better one.

The time from puberty to the middle or late twenties is the period when the sentient body and the sentient soul develop. This again makes it easy to see why the schizophrenic process initially affects mainly the sentient life of the middle sphere. Growing coldness in sentient life is a characteristic early symptom of schizophrenic psychosis, and in some cases the above-mentioned developmental disorder of the sentient body or sentient soul may combine with this. As the psychosis progresses, the congealing sentient responses grow distorted–as in the above case history–or they dissolve and are silted up. The awareness soul is also involved, in a sense prematurely, as is evident from an often excessive awareness in the above-mentioned 'demented struggle for conscious awareness'.

Schizophrenia only rarely occurs as a premature process in childhood, yet it is at that age that environmental factors play a major role in creating the disposition for schizophrenia. A difficult family situation is common, and the essential aspects of this have been discussed under the causes of developmental disorders; in schizophrenics, too, lack or excess of warmth in the family nest is pre-eminent. Another factor that has been mentioned by a number of authors is 'intellectual stimulation' in the families of schizophrenics.[45] Generally speaking, 'childhood misery gives rise to schizoid attitudes and these schizoid attitudes predispose to schizophrenia.'[46] Environment and disposition combine, and according to Bleuler 'action and reaction develop between inherited developmental tendencies and environment.'[47] The emphasis may be on either of the two components. The contact disorders of schizoid individuals that have developed on the basis of environmental factors lead to autism, or they may enhance an existing disposition for autism. On the other hand they may actually be the cause of the disposition becoming manifest. This applies to the whole field of schizophrenic psychosis and to other psychological conditions. In some patients–Hoelderlin for example–we may well think that the disease would not have developed at all, or only to a minor degree, if it had not been for the powerful influence of the environment. An inherited disposition is not an inescapable destiny–particularly if we take the view that the individual actually looked for it– but rather the material out of which individuals and their environment can shape their destinies.

This points to the responsibility that lies with the environment and particularly the education given at school. The negative factors due to wrong education have already been discussed. Significantly, schizophrenic patients will quite frequently have a history of having been model students and even top of the class at school. Relating this to the slender (leptosomatic) build, Kretschmer developed the concept of a 'leptosomatic record holder'. The emphasis is on memory and intellect, with the life of will and feeling more than usually in the background. Kretschmer wrote that modern secondary schools were one-sidedly geared to this constitutional type.[48] This type only comes fully to expression in the third seven-year period and has a particular affinity with the schizoid character type. Once again we are therefore able to say that modern education is a factor in the development of schizoid character traits.

In his first course of lectures to the medical profession Steiner said with great emphasis that the education given in the first and second seven-year periods could induce or prevent schizophrenia. It would be induced if teachers ignored the fact that education should encourage imitation in the first seven-year period and be based on authority in the second seven-year period.[49] Again it is possible to see that if children cannot vibrate in harmony sufficiently in their metabolism through imitation, and if they do not really learn to love the person who wields authority and strive to follow that person out of their rhythmical systems, they will not develop adequate foundations for the life of will and feeling in the lower and middle regions of the organism, and consequently fail to incarnate sufficiently in those regions.

If children become fixed in the head this causes creative powers to be excessively withdrawn from the head and leads to the development of an asthenic or leptosomatic constitution. The term 'asthenic' is used to stress the reduced vitality of this constitutional type which is on the increase in the present time. In exchange for an accelerated gain in height the internal organs fail to mature properly. If one considers the physical build it is immediately apparent that the ether body's powers of growth are drawn up towards the head and are only able to take the body along with them in longitudinal growth but not give it its proper roundness. An organism shaped like this, with destructive processes getting the upper hand prematurely, develops too much conscious awareness. The increasing nervousness seen in children, and particularly school children, may be connected with this type of constitution, or with reduced vitality in the life of a body that may be different in build but nevertheless is weakened and easily exhausted, i.e. too many demands are made on it by the nervous system. The neurasthenic type of illness described by Steiner, which is dominant today, here presents itself at the functional level, the level of the ether body, whereas in the above case it also comes to expression in an asthenic build.

Incomplete incarnation will evidently favour the excarnating process of a psychosis. The question as to why the tendency to emphasize head principles that exist in our time may give rise to obsessive-compulsive illness in one case, hysteria (due to 'flight' of the astral body) in another, and schizophrenia in a third is in the final instance a question of destiny. With regard to the physical basis we may say that the more profound the physical weakness, and the more difficult the incarnating process becomes, the more serious is the psychological or physical illness. With obsessive-compulsive and hysterial conditions the incarnating process comes to a halt in the middle region; with schizophrenia, the most severe of the psychotic disorders, even the middle region is only partly reached.

Proteins, the structural components of the body, must also be seen to be subject to different degrees of damage. In obsessive-compulsive illness protein synthesis is poor and there is a sulphur deficiency; in hysteria one has abnormal porousness; in schizophrenia finally the damage is most severe and takes the form of defects in organic protein. Life has not fully entered into the organs and during the first two seven-year periods weak spots develop; breakdown in these leads to the 'breaking out' of organic powers and hence psychosis. The breakdown itself is often caused by experiences individuals have been unable to assimilate at the conscious level so that they are shaken to the marrow. (In German the saying is that things 'go to the kidneys'. They will first of all reach the adrenal glands, the marrow or medulla of which excretes substances that allow things experienced in the soul to have a destructive effect on body proteins.)

Shocks, emotional or intellectual stress situations, i.e. situations causing increased tension in the astral body, thus become triggering factors for the onset of schizophrenic psychosis. Bleuler put particular emphasis on 'love gone wrong',[50] an experience that particularly affects the middle region. This played a significant role in the above case history (and with Hoelderlin). The disaster assumed such enormous proportions for the patient because he had hoped that his 'love' would give him some of the things that he had unconsciously missed in childhood.

Both the asthenic (leptosomatic) constitution and the schizoid personality only present themselves clearly after puberty. The leptosomatic constitution then holds on to the active growth phase that is physiological in the prepuberty stage, whilst the schizoid personality exaggerates and fixes the schismatic psychological phenomena of puberty. Increased intellectualism and emotionality, normally a transitional stage, determine the personality for the years to come.

The actual schizophrenic psychosis, its onset frequently marked by further aggravation of the problems of puberty, generally only becomes manifest after puberty and involves a pathological return to puberty. This has to do with the new situation that arises for the astral body when it loses not only the maternal cocoon as it is born but also the protection afforded by the physical body. Not fully incarnated and now 'left to itself',[51] the astral body succumbs to the polarities inherent in it that now develop into schisms. They in turn open the sluice gates for the creative powers in the physical body that have been withdrawn from the body through errors in education and now ascend further into soul life. So far it had been possible to transform them into powers of thought and of memory, but with psychosis this type of transformation is no longer possible. Withdrawn from the 'cosmic logic' of the organism,[52] not properly taken up into the logic of the (de-ranged) ego, the vital powers of the organism, having so far been misused, are given full scope in a soul that is now delivered up to them.

Here we perceive once again the two types of rebellion (described earlier), though they now take pathological form. One of them is the emotional rebellion of puberty. At the organic level it shows itself in the following way. As the astral body comes to birth, the urogenital system comes into play at the opposite pole to the brain, which until then has been dominant and used one-sidedly. The astral body, feeling exposed to the world as it comes free, draws on that system for powers of rebellion against the world, against all finished end-states, and in the final instance against the individual's own head. Schizoid individuals not infrequently feel themselves that the exaggerated busyness of the brain is a torment, and their frenzied rages are essentially directed against this. The second rebellion is organic, it is the rebellion of the creative etheric powers that goes hand in hand with the onset of schizophrenic psychosis.

This gives rise to the counter process, or counter image, to the intellectualism of our time, which we now perceive schizophrenia to be. Here you have the mad individual, whom society prefers to shut away, and there the individual who is top of the class, the perfect student, the child pushed into intellectualism. In quite a few cases it is one and the same individual whose sufferings not only cry out for help but should be a warning, should wake us up to consider the conditions under which human souls develop today and in the future. (As already mentioned, there is a danger that at some point in the future schizophrenia may become epidemic. At the present time about one per cent of the population is affected by this, the most severe and most frequent of 'mental' illnesses.)

The beginnings of schizophrenic psychosis can be seen not merely in individuals but also in the whole of western humanity. We discover the schism between soul and world that we have spoken of and which is the cause of the inner isolation of modern people. We see the schism between intellectualism and emotionalism, between perception and action. Something the head recognizes to be the right thing to do does not come to realization, and sometimes subjective and emotional motives make people do the opposite. At the level of humanity we see 'too much head'; in individuals this is exaggerated to become a schizoid personality and the disease known as 'schizophrenia'.

Excarnation and Regression

The last-mentioned aspect–schizophrenic psychosis as a counter process to the intellectualism of our time–brings us to the subject of this chapter. The counter process goes hand in hand with the regression that has been described. Yet how does this regression combine with the tendency to excarnate? The above title–Excarnation and Regression, which has relevance for all psychological conditions–would seem to be paradoxical, for excarnation, going in the direction of death, surely means anticipation rather than regression.

We have in fact noted such an element of anticipation at certain stages in all psychological conditions. The ageing process of obsessive-compulsive illness, the leaking astral body in childhood hysteria, with anticipation of puberty, the epileptic's tendency to leave the body between convulsive attacks, the schizophrenic's derangement relative to the body–all this points to partial anticipation of later life stages and of death. It will be seen that depressions also show this trend.

In obsessive patients and in schizophrenics the asthenic constitution with its emphasis on the nervous system and its low vitality showed characteristics of old age. In obsessive-compulsive illness, life is pathologically determined by a thought life based on the forming of images and on ordering principles that is normal in older people. In schizophrenic psychosis, remembered images are often assimilated in a pathological way; these are soul contents that are perfectly healthy in the old. We shall find something similar in the case of depressions.

The situation is different, however, if we consider the dynamics of psychological illnesses. In obsessive-compulsive illness (and depression) they are largely determined by the head and lead above all to anticipation. In other psychological conditions the lower pole of the organism and of soul life plays a greater role, causing the dissolution of previously established soul structures and regression. Adult hysteria partly repeats the years of puberty. In the patient suffering from severe epilepsy we perceived the 'structure of an adult child' which may also only develop in the course of life. Schizophrenia took us back to the emotional upheavals and the schisms of puberty; it may sometimes lead to a second, pathological childhood which will be discussed under the heading of therapy. Severe obsessive-compulsive illness or depression may also cause patients to fall back into the helpless state of a child. Mania will be seen to present a picture of a return to youthful vigour.

We have seen that development can never stand still in the biography; it will either go forward or regress. On the other hand it may be subject to schism, with progressive development coming from the head, whilst from the lower pole the soul seeks to regain its youth. The same happens with psychological conditions, where the emphasis may be more to one side or the other. The tendency to excarnate is countered when in the partly excarnated state a tendency to regress to childhood develops. These reflections on 'Excarnation and Regression' started from a consideration of schizophrenia, and it is particularly with this disease that new incarnation processes occur in a soul life that is no longer firmly engaged. The organic powers that have ascended from the ageing body cause new organs to grow in the life of the soul, and the deranged ego unites again with these. Putting it succinctly we may say that the patient dies back into a pathological childhood.

We may ask ourselves if this life of regression that occurs in conjunction with the partial dying process of psychosis has any connection with the retrograde review of past life that occurs in the purification phase after death. This has been described by Steiner (e.g. in his *Occult Science*) and is known as 'purgatory' in religious contexts. Some schizophrenics do indeed go through a kind of purgatory, when past events that have so far been harboured in the soul come to them anew. This experience of past events arises because the soul turns inside out, as already described, something that happens in a more comprehensive way after death. Compared to the time after death, the partly excarnated schizophrenic patient does not come free of the body and therefore is unable to achieve the total picture gained in freedom and the conscious purification that the after-death experience is able to give.

Regressive experience thus becomes regressive life in schizophrenics; it does not free them so that they may ascend to the world of the spirit, but demands a new, healthy incarnation in the earthly world that must be achieved through healing of the illness. The tendency to regress that develops with the psychological illness must be regarded to be due to the fact that the patient remains bound to the body; this state is maintained in spite of the excarnating tendency and may indeed be enhanced if the right cohesion between the principles is lost to an even greater extent.

With this, we return to the subject of schizophrenic psychosis as such. The next question to be considered is how this psychosis relates to the supernatural world.

Schizophrenia and the Supernatural World

Schizophrenics speak of supernatural experiences, of encounters with demons, angels, and with Christ. The patient whose history was given above also saw a Christ figure in the clouds who spoke with a voice of thunder, and this was accompanied by flashes of lightning and by light phenomena. In such a case as this we may think beyond the creative powers of the ether body to fundamental powers inherent in it that create light and sound phenomena out of the patient's memories and inner experiences. Such a power, the light ether, is released from the kidneys. Steiner spoke of 'kidney lights'.[25] [The German term he created for this is analogous to the word for heat or sheet lightning and conveys that image–Translator.] The voices that are heard call to mind the powers of the sound ether that have their centre in the liver. Does this mean that these experiences merely indicate the release of powers in the inner organism and that they have no reality beyond this?

When schizophrenics report supernatural experiences we may ask ourselves if something of the supernatural world is not showing through that takes its first impetus from memories and may be clouded and distorted by the illness but nevertheless suggests a spiritual reality. We come to realize that the patient's ego, astral body and ether body do not merely exit from the physical body and the world we perceive through the senses, but that they also enter into a new, supernatural world. This, however, is nothing new to them. A new aspect opens up in our investigation: Human beings belong to an etheric world through their ether bodies, to an astral world through their astral bodies, and to a world of spirit through their egos; the higher principles maintain their link with those worlds throughout life. This link is a much more direct one when those principles are active in the physical body and have not yet come to themselves and to birth; later, when the principles are largely devoted to the soul life of the individual, it is more indirect, and this creates the opportunity for establishing a new and conscious relationship to the supernatural worlds. Certain aspects of the direct relationship are repeated in psychosis as the powers of ether and astral body and of ego bring the original relationship to the supernatural worlds back to life in the soul. This revival does however take place in a soul that has now achieved conscious awareness, with the result that elements that formerly had an unconscious influence on the child's life now come into the light of awareness. Because of the illness this is a chaotic process.

Steiner spoke of 'mentally ill' people who spoke nothing but gibberish while ill. After their recovery they told of perceptions made that arose in the world of the spirit; again they produced much that was irrational, 'but also much that was right'. When their brains were functioning properly again, he said, they reported the out-of-body experiences they had had in the world of the spirit during their illness.[53]

Such a situation also arises in schizophrenia. With other schizophrenic patients, the above patient, for instance, one has the impression that some of the supernatural experiences come to awareness even whilst they are caught up in their illness. Something that in others only shows itself when the illness comes to an end flashes up during the intervals in their case. It usually takes a pathological form, however, unless the patient is able to wrest from the disease a healthy form of it that has objective validity, as Hoelderlin was able to do for a time in his artistic work.

People suffering from psychosis are thus able to cross the threshold to the supernatural world, for the connection with the physical body is partly broken for a time, as it is done in spiritual training. The British psychiatrist R. D. Laing has recently referred to the way the mentally ill cross the threshold. In his view, certain transcendental experiences appear to be the origin of all religion, and some psychotics also have transcendental experiences. They may be illumined by light from other worlds that may also burn them out. Laing posed the question as to whether the trips into the supernatural reported by schizophrenics should in fact be called pathological. Are they not a 'natural road' to recovery from the dreadful state of alienation that is called 'normal'? Psychotics who are illumined by the light of other worlds encounter 'demons and spirits' and make this known to us,[54]

In Laing's experience, supernatural worlds shone through into the earthly world of today, worlds that modern people no longer want merely to believe in, but want to experience. According to him, schizophrenics are able to convey experiences from those worlds. Considering what has been said above, this seems possible within limits, though in the long run it does not help psychiatrists or their patients. Patients can make their psychiatrists aware of the supernatural sphere, and they will feel different and in fact only really understood when they find that the physician has a positive attitude to the world of the supernatural. They then also want to be cured in the right way, i.e. not merely cut off from their supernatural worlds and immured in a standard of health that is alien to them. Laing therefore also calls for an approach to treatment that is different from the one that is generally followed today.

Such an approach to treatment, where the development of soul and spirit is taken into account and the physician is also able to follow patients through the 'stormy passages' of their journey into supernatural worlds, will however only be possible if the perceptiveness and the strength this requires are gained by crossing the threshold in a healthy way. Steiner crossed the threshold in this way and he has shown the way to do it. Before we go into this in the last chapter, it will be necessary to consider the fundamental difference between the healthy and the pathological way of crossing the threshold.

The preconditions for crossing the threshold in the course of spiritual training are increased health of soul, and increased inner activity and powers of judgement. The precondition for a psychotic crossing of the threshold on the other hand is psychological illness, with supernatural experiences distorted as that illness bursts in upon a passive soul, and no longer clearly discerned and judged by the ego. This becomes evident if we compare hallucinations and perception through images. (Concerning the latter, see the final chapter in this book.) Steiner summarized the difference as follows: 'If the body as body creates images: hallucination. If the spirit as spirit creates images: perception through images.' Image-forming life at the unconscious level creates the living physical body. 'Progression to higher vision arises if we partly withdraw ... from such activity in the physical body and raise elements that normally bubble and boil down below in the physical body to the level of conscious awareness.'[40]

The creative powers of the ether body thus play a role in the creation of organs of soul and spirit that lead to perception through images. Some of the powers that the soul uses (through the mediation of the powers of the ether body) to restore the body during sleep are used to build up the new organs for the spirit.[41] These organs may be compared to the sequences of images that become organs of hallucination and delusion in pathological soul life. The distinction is as follows: The creative etheric powers that are involved in the development of spiritual organs do not derive from a pathological outflowing in the physical sphere, whereas the creative powers of pathological organs do. They derive from an enhancement of healthy processes of transformation that otherwise act from the ether body into the soul and serve the thought processes. In psychosis we have an incomplete metamorphosis, as creative powers that should relate to the body become active in the sphere of the soul; a complete metamorphosis is achieved when perception in images is actively practised and the creative powers of life do not merely die in the thinking conscious mind but come to be resurrected in an expanded conscious awareness. The 'etheric token images'[55] that arise are revelations not of the body but of the spirit.

Yet even now, with thinking enhanced to become vision, the disciple must always be aware that these images are not the reality but are indeed images. Such awareness can however only arise if the faculty of judgement is enhanced, just as activity had to be enhanced to achieve the new metamorphosis of the creative powers. Both of these are beyond the range of possibility for the psychologically ill.

This account of their genesis clearly shows the main characteristic that distinguishes perception through images from hallucination. Hallucination, produced by the body, has the above-mentioned characteristic of being more or less physical; perception through images, produced by the spirit, does not have this characteristic. With any kind of spiritual vision, Steiner's words apply that he spoke when he described the 'colours' seen in the 'aura' of human individuals by a spiritual scientist. 'When clairvoyant perception speaks of seeing red this means: I have an experience in soul and spirit that is the same as the physical experience gained on perceiving the colour red.'[56] This also applies to other contents of spiritual vision, for these are not perceived in the physical sense, like hallucinations, but in their spiritual essence. Again one would not expect psychotics to be able to judge this.

Having achieved transcendental experience, the disciple fully turns his attention to earthly life again, knowing himself to be deeply committed to it–in which he differs profoundly from a psychotic subject. Unlike the sick individual, he also gains new constructive powers from the cosmos that will benefit him and the world. The final difference thus arises in the social sphere. Crossing the threshold in a healthy way tends to bear fruit for the life of the individual and for the world. Crossing the threshold because of psychosis, on the other hand, causes disruptions in the patient's own life and in the world that in extreme cases lead to elimination of all the social contacts of adult life and a pathological, chaotic recurrence of the situation known in youth and childhood. Any crossing of the threshold due to psychosis initially involves regression in soul and spirit, and it is only in exceptional cases that the patient is able to metamorphose this regression into progress. Husemann described one such case, where finally the question arose as to whether it was schizophrenia or initiation.[57] The majority of patients for whom such a metamorphosis becomes possible will therefore need help from their therapists. (See the chapter on *Therapeutic Aspects*.)

Healthy development of the soul gradually frees soul and ego from the physical body on its own. This culminates in the path of spiritual training, where individuals are at times able to rise out of the body and into the world of the spirit. In psychosis the disease lifts soul and ego out of the body. This does not really free them of the body; instead, they are pursued by physical forces that enter into them. Schizophrenic psychosis in particular may thus be seen as the pathological counter image of healthy soul development and over and beyond that of the path of spiritual training. An individual who succeeds in taking even the first steps on such a path, so that soul development is intensified and taken further, will also be taking important steps in the prevention of schizophrenic psychosis. Treatment has to take another course, however, and this will be shown in the next chapter.[58]

Therapeutic Aspects

As in the case of epilepsy, the medical treatment and remedial eurythmy offered by anthroposophical medicine are the most helpful. For remedial eurythmy, see the chapter on *Art Therapy*; the exercises based on the sounds 'A' (as in father) and 'B' are particularly relevant to schizophrenic psychosis.

As to the psychotropic drugs that are generally used today, essentially the same applies as with antiepileptic drugs. Psychotropic drugs, too, only have an effect on the symptoms and not on the disease, which may 'go underground' and break out again once the medication is discontinued. There is also a danger that the suppression of symptoms such as hallucinations and delusions may cause transformation of the pathological process into a milder but progressively chronic form. True cures have however been recorded with psychotropic drugs. In those cases the powers of self healing were clearly strong enough to come to terms with the pathological process once the symptoms had been removed.

Most schizophrenics will be calmer, more relaxed and adaptable if given psychotropic drugs, yet at the same time they also become more unproductive. Their reactions are less subtle and lack life; after prolonged medication they grow dull and apathetic. Discerning patients who had been on psychotropic drugs have frequently said that it felt like being 'walled-in' in the soul, and this may serve to sum up the action of those drugs in the present frame of reference. The abnormalities relating to the blood, liver, brain and heart that are sometimes reported as 'side effects' of psychotropic drugs form the basis of this limitation that is imposed on the life of the soul.

The sensation of being walled-in also shows that there is a certain positive aspect to these drugs, and in anthroposophically orientated medicine we are as yet unable to do without them in very serious cases. To give an analogy: When the spinal column threatens to fuse in a case of tuberculosis of the bones, temporary immobilization of the spine by means of a plaster cast may be required to prevent irreversible damage to the spinal cord. In the same way, a psychosis taking a rapid or vehement course may call for the soul to be walled-in because there is a threat of soul structures irreversibly dissolving, damage that so far cannot be prevented by giving the slower-acting anthroposophical medicines on their own. Wherever possible, psychotropic drugs (like the plaster cast) should however only be used to help the patient through an emergency situation that may be of shorter or longer duration. By using these drugs in combination with the actual treatment (which of course is also being given whilst the spine is immobilized), it is usually possible to manage with relatively small doses. Relatively severe chronic schizophrenias showing good recovery have so far often needed such small and very small doses for long periods, to enable the patients to keep their hold on life.

With small doses, the benefits outweigh the harm that psychotropic drugs inevitably cause. It is of paramount importance to know what one is doing when psychotropic drugs are prescribed–and never to let this rapid 'therapy' that on the other hand is self-defeating and paralyses the soul paralyse oneself so that one ceases to develop new therapeutic initiative.

The help given by psychiatric methods used intensely and by work therapy is particularly important in the treatment of schizophrenia, and in severe cases indispensable.

Modelling is the best form of art therapy in all cases where the emphasis is on dissolution and chaos at the level of the body or of the soul. Here it is even more important than with hysteria to help to 'seal' the organs by appropriate modelling and pottery work. Painting stimulates sentient life where it has grown weak or frozen; care must be taken, however, to bring a strong form element into work with flowing colours, otherwise new impetus may be given to dissolution in soul and body. Music has a specific role in schizophrenia in so far as patients may come to experience how strict mathematical laws that can be determined by using the rational mind are harmoniously united with direct experience in the life of feeling. In music, patients are able to sense and gain something of the oneness that was lost to them in childhood. Therapeutic exercises may start with dissonances that are in accord with the patient's state of soul and progress to consonance. Patients are particularly helped to overcome the schism between themselves and the world through therapeutic speech exercises.

The principal aspects of therapy for the spirit have already been discussed in that section of the general part. The advice given there, to be cautious in conveying spiritual contents to psychologicaly ill patients, applies particularly to schizophrenic patients. The same is true for indirect spiritual therapy: the effect of the image of the schizophrenic patient that the therapist has formed and is nurturing in soul and spirit is particularly powerful in cases of schizophrenic psychosis. Once the patient's ego can be directly addressed, spiritual contents may be conveyed again, though with due caution.

Patients who have recovered sometimes show a more direct relationship to the spiritual sphere than other people. When the delusion has been overcome, the organ for truth may be all the more well developed. Following recovery, individuals are able to relate better to the environment and have a greater awareness of inward unity of soul than was previously possible. Here one perceives something of the meaning of the illness.

Again it is apparent that therapists gain a great deal from intense study and reflection on the laws of biography. Thus one comes to realize that schizophrenics show a partial regression to the third seven-year period. It is therefore necessary to be cautious in presenting spiritual contents because the ego is not fully present, which corresponds to the situation that pertains in the third seven-year period. On the other hand it is possible to appeal to the patients' struggling power of judgement, e.g. by discussing a particular delusion with them.

If regression has taken the individual back to the second seven-year period the route via logical judgement can no longer be taken. According to Steiner, repeated confrontation with reality can be helpful to these patients, 'particularly in early youth'.[59] One may for example confront a patient with someone he thinks is persecuting him and who is able to show the patient that he cannot possibly have been persecuting him at the time, because he was somewhere else. Such things may also be conveyed in an authoritative manner, for the right kind of authority wielded (with love) by the therapist can be a tremendous help. Patients often long for this kind of authority at this stage of the disease, particularly if they have not had it in childhood.

If schizophrenic patients have partly regressed to the phase of imitation, i.e. the first seven-year period, and are for example constantly repeating the gestures and words of others, a family atmosphere should be created around them where their souls are encouraged to vibrate healthily in tune with the adult world. They should be able to take in some of the warmth of the family nest that was lacking in childhood. Again it is important always to remember that above this second childhood in its pathological distortion there is the patient's healthy ego and that this is the goal to which new development wants to ascend out of the illness.

Depression and Mania

Clinical Picture
The wife of a senior civil servant came to see me. She was 32 years old but looked older than that. Of stocky, short and thick but formed-

out build (pyknic type), she moved slowly and with seeming effort. The face was round and lax, and facial expression was limited, and showed profound sorrow. She was reluctant to raise her eyes, and the voice was low and lacking in strength. Her thinking was as slow and heavy as her gait. She would sometimes fall into silence whilst speaking, becoming completely immersed in herself, and then continue, omitting elements in her line of thought. Occasionally there would be a glimpse of warmth in her eyes, or a warmhearted word would break the darkness of soul, for instance when she remarked that surely it could not be easy to have to listen all day long to the troubles of so many people.

The patient said she had come because of a feeling of pressure on the heart. Her heart had been examined a number of times and she had been told there was no organic disease and that she should go and see a nerve specialist as she was probably suffering from a cardiac neurasthenia that was psychological in origin. Asked what it was that lay so heavy on her soul, she said, in an even lower voice than before, that (for about eighteen months now) she could no longer cope with her responsibilities as housewife, wife and mother. It was particularly bad in the mornings, when she could hardly manage to get up because she was so afraid of the day that lay ahead. Throughout the day she had to make a special effort before she could tackle even the smallest things. More and more of her work was not getting done. She also could no longer be the woman her husband needed, nor could she meet the needs of her children. She always felt so useless. Even when her husband had exercised his authority she had not made enough of an effort to overcome her laziness. Quite the contrary: she had then stayed in bed altogether.

Asked if she had ever felt tired of life she admitted that she had. The question as to whether she had actually thought of doing away with herself was only answered in the affirmative after quite a long period of hesitation. She denied, however, that this had led to concrete proposals, saying that this would be unfair to the family. But then she followed this with: In the long run it would probably be a relief to them if she was no longer there. She was a failure, she had really always been a failure, though she was only now coming to realize it. And surely that was another bad thing, that she had only found this out now!

From the history it emerged that her grandmother had suffered from depressions and had been admitted to a neuropsychiatric unit on a number of occasions. Unlike herself, however, her grandmother had been really ill.

Her childhood had been harmonious. She had always been very much attached to her parents; their family life had been so wonderful. Detailed questions elicited the following. She had been quiet as a child, but always sociable and usually happy. She had always had a tendency to take things to heart too much and found it difficult to cope with sad events. She would then be withdrawn for days, and low in spirits and barely able to respond, particularly in the mornings. But of course it had all been her own fault. Sometimes she also could not cope with joyful events and would be brooding over them for a long time. She really wanted to 'understand' things all the time and know how they had come about. When such a period had passed she would be cheerful again and sometimes even in really high spirits. She would then be utterly carefree, spending too much money, buying anything that took her fancy. Later she would feel great remorse over this, and surely that was only right.

Her father had died in a traffic accident when she was eighteen. That had been a terrible thing for her, but because she had to help her mother she had got over it in a few weeks. (According to her mother, who had been in a depression at the time, she had been quite desperate at the time, especially in the mornings, but had nevertheless really taken her mother's place for a time and been very brave.) She had married when she was twenty, a senior civil servant from the Ministry of Justice who was fifteen years older than herself. She had felt so unprotected since her father died, and so she had accepted his proposal. And the marriage had turned out well. Her husband was utterly dedicated to his work and always bringing papers home, but she had come to terms with this. Her two children, who had been born when she was twenty-one and twenty-two years old, were a source of great happiness to her. She was able to give to them some of the things that had been given to her when she was a child. Since they had started going to school she was no longer needed the way she had been needed before, however. This had been hard for her. And it was since then that she felt this pressure on her heart that had made her come to the surgery.

Physical symptoms mentioned by the patient were digestive problems: A slight sensation of fulness after meals, intolerance of rich foods, wind, and constipation. An illness that was of interest in the given context was jaundice at the age of twenty-four. She had had the same digestive problems for years before that, but they had been much worse then, whereas now she hardly noticed them.

The Pathological Process

The patient showed a certain hereditary trait, which is not uncommon with this type of depression. The environment initially showed nothing that would have a untoward effect; there was harmony at home. Problems arose from inside, because she was inclined take things the hard way. Even less serious and indeed joyful events were not easily coped with. They sank down and kept coming back, acting as foreign bodies in her unconscious and then also conscious soul life that could not be assimilated or 'digested' properly. The patient wanted to 'understand' the things that happened, before they came to be forgotten, and this proved difficult. Here the hypersensitivity to environmental factors shown by obsessive and hysterical patients, which had gone inward in schizophrenics, had become a problem of coping with events inwardly. Withdrawal was in this case combined with insensitivity towards the environment at times, something we have seen in a more acute form in epilepsy.

Depressive moods with sadness began to develop in response to certain events. They would terminate either in a balanced experience in heart and soul or in the opposite mood (only a mild form of it in the case of this patient) that we would call a hypomanic (i.e. slightly manic) phase. (Talking to the mother it became evident that the patient was exaggerating when she called herself 'utterly carefree' at this stage.) Unlike the full-blown manic state this was merely an extended period of somewhat high spirits that arose when the melancholia had resolved.

Up to a certain point, sadness and sorrow are part of normal human life. When we feel sad we are given opportunity to come to ourselves again after living too much in superficialities for a time. We are given the opportunity to reflect and to gain new insights, and new powers of will, by overcoming sorrow. Melancholia is therefore part of us, and so are times of high spirits. The path of development for the intellectual and mind soul lies between melancholia and high spirits.

People who are particularly prone to such variations in mood are called cycloid or cyclothymic to distinguish them from schizoid types in modern psychiatry. The syllable 'thym' (Greek *thymos*) refers to mind and spirit, and these are subject to cyclic variation. Compared to the airy element that unleashes the storms of excitement in a schizophrenic soul life, we here experience the watery element as the determiner of soul life. This element does not tend towards dissolution in the soul, the way the airy element does, but is apt to create fluctuations in soul life, phases that keep returning to a middle position, like the waves of the sea.

At the organic level this immediately points to the liver as the central organ of the water organism. If soul life has its foundations in this organ we get the phlegmatic temperament of the liver person, where the slow circling motion of the water organism becomes a ponderous movement caught up in itself at soul level. With reference to the higher principles this is the sphere of the ether body, where temperament and mood have their roots. As already noted, the intellectual or mind soul arises out of a transformation that occurs in this sphere.

That is also the soul aspect we experienced in the patient. There is a marked tendency to assimilate via the rational mind, to take things in through a thoughtful mind, but then the assimilating ego activity fails to get a real grasp and the process comes to a halt as things are taken too much to heart. We experience the element of gravity, which is connected not with the watery but the solid element. We note deviation from the middle position in a direction that in schizophrenic patients led to a frozen state lacking in feeling and will capacity, but in this case produces quite a different picture, a picture we are able to follow in heart and soul.

The image of a melancholic soul that was presented by the above patient is defined by the diagnosis of melancholia or endogenous depression, i.e. depression arising from inside, in psychiatric practice. In this case the soul does not react to an event that would cause sadness, and such an event has no or at least only a triggering role; the illness arises from inside the soul. The diagnosis 'cardiac neurasthenia' is not correct at this stage as the process is a psychotic one and the heart is only secondarily involved.

In the general discussion mention was made of the principle of gravity that relates to the solid element. Here we have immediate experience of it. Basing itself on the lung it causes a melancholic temperament to develop, with the liver only involved secondarily. Basing itself on the liver it causes the condition known as 'melancholia' to arise, with the melancholic temperament going wrong at the level of the liver. Gravity, weight, now drags down the limbs, the facial expression, the eyes and the thoughts, so that a line of thought can no longer be followed in all its stages and ideas and words are omitted.

In the sphere of movement, paralysis takes a particular direction. Movement due to the astral body intervening involuntarily is not affected; emotions such as fear may even lead to accelerated reactive movements, and inner unrest may cause marked physical restlessness. Intentional, will-directed movement originating in the ego is inhibited. The patient could not carry out the actions she intended, because the movement of her will rather than her limbs was inhibited. With the 'best will' in the world she could not bring her intentions to realization.

Paralysis of the will, or at least inhibition of the will, is the central symptom, with all other symptoms arising from it. This paralysis leads to a feeling of being paralyzed and inhibited in the soul; this 'experience of lifelessness' in the soul[60] then gives rise to profound sorrow and the feeling of utter uselessness that the patient spoke of. We can also see that depressive patients get stuck in the past because of this paralysis of the will, and keep going over and over everything they have done wrong.

Surely we all of us have the beginnings of this type of depression in us. What do we find, as we look back into the past and judge the way we acted and behaved? If we are honest we have to admit that much has been left undone and many faults were committed. If we were to stop at this point we would have good reason to become depressed. Something prevents this, however, and that is the knowledge and hope that we shall make further progress in the future, be able to perfect our imperfections and make up for our faults. Resolving to do so, we first of all build a bridge for the will that goes from past to future in our thinking, and in our feelings, which reflect the present, hope arises and gives lift to the arching bridge of the will.

This is something patients suffering from endogenous depression are unable to do. The will that should take them into the future and bring the future to realization is paralyzed, and they no longer have any relationship to the future. They are hopelessly caught up in the past and feel that every imperfection, every fault, is final; on this basis the future can only be experienced as a continuation of the past, or a repetition of what came to pass before, with all its imperfections and faults. The future thus can offer only further demands that can never be met, and these patients can only fear that future. Depressive patients thus develop a fear of life. Their conscience experiences every personal recollection of the past as a self accusation that weighs on their hearts.

This situation, from which there seems to be no way out, also explains why the urge to commit suicide is sometimes irresistible. From this point of view it is a desperate attempt to put an end to an existence without goal and purpose that is a burden to others, and to do so here and now, as the only realization of the future of which these patients may feel capable. Yet even this 'future' is entirely determined by their experience of the past and does not rouse any hopes.

The Manic Phase

The manic phase, which may be followed by a depressive phase, presents more or less the exact opposite of depression. It is also much rarer than the latter and will therefore only be given brief consideration.

If it is more than a reactive hypomanic phase of the kind seen in the above case history, the clinical picture is as follows: The patients move quickly, their feet are winged, and they are always seeking to escape gravity; instead of paralysis of will they experience great will impetus. Instead of the dark sorrows of depressive patients we perceive the sparkling fire of the will in their eyes. This fire also gives thrust to their speech and flashes forth from the life of the soul; the urge to bring the will to realization is constant.

Given wings by spirits that are raised 'from below', patients develop an intense desire for life rather than fear of life. They seek to enter into life with avidity, to ge rid of their powers of will in actions, and these are actions where they give in to any wish or sudden desire that comes to mind. Depressive patients are no longer able to relate to the future; manic patients with their excessive will element are one-sidedly given up to the future. They have lost their connection with the past, sweeping aside past experience in the fire of their desire to act. Depressive patients look old; a patient in this type of manic state seems younger in both body and soul.[61]

It is apparent that in place of the shift towards the solid element seen in depression there is a shift from the watery middle of a liver-based soul life to the pole of the fiery element. The warm and placid phlegmatic liver temperament, which on the one hand may incline towards depression and melancholia, has become a pathological, abnormal choleric temperament with emphasis on the process of will. This may culminate in attacks of frenzied rage as an extreme form of excitement that may also develop in schizophrenic patients. In manic subjects the sanguine temperament of the air organism may also be involved, so that manic patients may be rather airy-fairy people. The central element, however, is always the fire element of the will.

Organic Aspects
Again a foreign element that cannot be understood solely in terms of the soul has come up in the life of the soul. In the final instance this foreign element arises from inside, but from below, from the physical body; this is also the view taken in conventional psychiatry. Research has shown that (as in the case of mania) depression goes hand in hand with a number of metabolic disorders, only brief reference to which is possible at this point. They culminate in congestion in the water organism where certain substances are again retained, more or less as in the case of epilepsy but not to the same extent. The ancient Greeks were largely thinking of abnormalities in the flow of bile, and the word 'melancholia', which is still used today to describe endogenous depression, originated through this. (Stagnation causes the bile to turn black, and the Greek for black is *melas*.) Compared to liver metabolism, biliary metabolism has been given even less consideration in endogenous depression. On the other hand it has been established that there is a connection between worries and a slowing down in the flow of bile.[62]

Modern research tends to look for causes in the brain, yet a more real understanding is gained by considering the liver, in so far as it is an instrument of the soul. We have already shown it to be a foundation for the life of mind and heart that arises as the ego intervenes and causes transformation in the ether body. The liver is here addressing itself to the blood and to the heart; the ego lets the ability to assimilate in mind and heart arise, i.e. the desire to 'understand' how something has come about. As the liver addresses itself to the lower sphere, a second aspect develops: The biliary system which extends throughout the liver and opens into the intestine makes the liver the organ of the will, with biliary activity preparing for the internal combustion process of any act of will.[63] The will, however, enables us not only to move our limbs, words and thoughts, but also to move experiences that are to be assimilated in a reflective mind. Thus the mind and the will come together in the system of liver and bile, and we begin to understand what Steiner meant when he said: 'We are unable to feel properly if the liver is sick.'[53]

If stagnation arises in the flowing water organism and matter condenses in the liver system, in the flow of bile, the ego is no longer able to intervene properly in this system to set an act of will in motion, and the result is paralysis of the will and the emotional illness known as 'depression'. The stagnation has its origin in the sphere of the head, the one-sidedly paralyzing tendency of which has penetrated to the inner metabolism (which in this case is fully developed). Mind and will are shackled to images of what has come to be, to the past life of the ego, and are initially unable to free themselves from this bondage to the organic sphere. Compared to the mental images of obsessive-compulsive states, it is impossible to gain distance and hence insight into the disease in this case.

The question as to how such abnormalities arise in the liver and biliary system takes us back from the physical level to which we had gone down after considering depression at the soul level and return to the soul level, where physical liver disorders have major roots. The patient spoke of problems she had not been able to assimilate and how these weighed on her soul and also 'lay heavy on her stomach' in physical terms. Digestive disorders developed that pointed from the stomach to the liver. Difficult things that happen to people can put such a burden on the liver–particularly if the disposition is there–that jaundice develops for psychological reasons.[62] The foreign-body effect of such an unassimilated event acts via the ether body to produce stagnation in the water organism of the liver. If other factors are involved as well, this may even take the form of inflammation (hepatitis) and gallstone formation.

When this type of organic disease has entered a milder, chronic stage, or if the organic disorder was mild from the beginning, the pathology may in a small number of these patients have repercussions in the life of the soul. The liver 'gets used to irregular intervention from the astral body' (which carries the unassimilated events to the liver). 'This only needs to go on long enough, and it will take the opposite road into the soul sphere; the liver pushes elements that it should have assimilated at the physical level into the soul sphere and the result is depression.'[64] Minor congestion in liver metabolism brings heaviness into the life of the soul and leads to the severe paralysis of soul one sees in endogenous depression. It is evident from the 'discharge' phenomenon of the manic phase, which develops in relatively few cases, taking the place of a healthy resolution of the depression, that congestion lies at the root of it. Steiner spoke of such a discharge in hysterical children who in their soreness of soul have withdrawn into a state of depression. This of course is not the endogenous depression of adults which arises from inside. In both cases, however, discharge results in a 'manic state'. (For the partial reversal of metabolic disorders in cases of mania, see Reference 1.)

The question as to why these disorders have such a serious effect on the soul when they have become physically less acute, and why only a small proportion of liver disorders, which are very common today, lead to endogenous depression, has a first-line answer in the patient's disposition. Many patients suffering from endogenous depression have that trait of taking things hard in their histories; here the gravity principle of the body ascended into soul life and formed the character even before the psychological illness developed. If we are serious about the living relationship between body and soul and expand the concept of disposition, the following image arises: A soul life that takes things hard does so not only with regard to what it takes in of the world (and this then burdens the liver), but also to the relatively minor problems that come to it from the liver. Letting things weigh on one's mind thus leads via the liver to endogenous depression. For others, who do not have that disposition, minor liver disorders may have no effect on the soul at all, or they will at most produce a bad mood for which there is no apparent reason. From this point of view, moods like this may also be taken to be the beginnings of an endogenous depression.

Endogenous mood changes occur mainly in the mornings, and above all immediately on waking. That is also the time when endogenous depression is most severe, both subjectively and objectively. This characteristic diurnal variation in endogenous depression again refers to the liver. Morning aggravation is in many cases preceded by a painful awakening at about 3 a.m., with the patient unable to get properly back to sleep. This gives rise to a darkness of mind that has its roots in the paralysis of will and extends far into the day, sometimes only giving way to a lightening of mood and activation of the soul by evening. The darkness in the soul arises from a physical darkness. Metabolic studies on patients suffering from endogenous depression suggest that the night phase of the organism extends partly into the day, in so far as the depression of secretory functions that is normal during the night continues into the day. At night, this means rest and recuperation for the liver; continued into the day it means stagnation and paralysis, and further catabolism, destructive activity due to the day's events, brings pain in that situation.

Secretory functions should be reactivated or enhanced at 3 a.m. in every individual. This marks the gradual beginning of the waking-up process and also makes sleep more superficial. The ego and the astral body, having separated from the living body during the night, gradually return to it from 3 a.m. onwards, i.e. they first of all enter into metabolism via the limbs, among other things setting secretory activity in motion in the liver and establishing a will link with the world once the individual has come fully awake.[65] If they encounter an obstacle in metabolism, in this case a congested liver, the gradual wakening process becomes a sudden one. If organic liver disease is present the patient may wake with pain in the liver; the more subtle liver disorders of endogenous depression on the other hand cause pain in the soul.

After such a pathological awakening the ego and astral body remain more or less caught up in the liver system; in some instances they need the whole day to come to terms with the obstacle and only over-come it to some extent by evening. Analogous to epilepsy, it is difficult for them to penetrate the organ, but in this case they are not held imprisoned at the same depth in the organ, nor do they struggle physically against the obstacle the way an epileptic does in a seizure. The darkness does not extend to conscious awareness and cloud it the way it does in epileptics; it remains in the sphere of the will, casting its shadows on to mind and soul.

It is evident why therapeutic sleep deprivation, with the patient kept awake and active for a whole night, has given good results. The ego is kept engaged beyond the more favourable evening phase, and this can also give impetus to liver metabolism

The Illness in Relation to Biography
The difference from the situation that pertains with epilepsy becomes even more apparent when we consider endogenous depression in rela-tion to biography. There are childhood forms that represent preliminary stages; the mood changes seen in around the ninth year may arise because environmental factors reveal a disposition. These childhood forms only rarely develop into the illness that is seen in adults, however. This may occur from puberty onwards, though onset is most common from the age of twenty-nine up to advanced old age,

with a peak in the fifth decade. Alternation between depression and mania, i.e. cyclothymia, most frequently has its onset in the third decade.[66]

That is the period of initial preparation and also the period when the intellectual or mind soul develops. This stage of soul development may never reach completion in these patients. Mood swings, fixation of the rational mind on the past, and darkness of mind and soul are indicative of problems in the development of the intellectual or mind soul. Once again the heart, as the organ through which this soul principle comes to development, is centrally involved, though not, as in cardiac neurasthenia, by heart powers becoming retroactive, but by the soul pathology of depression. This connection with the heart is an essential element in endogenous depression.

Central involvement of the heart comes to expression in the tendency to develop guilt feelings, which is under organic compulsion, and to self accusation, i.e. accusations the self, the ego of the heart, makes against itself. This may may even take the form of delusions: 'I am the worst housekeeper there has ever been; I am not fit to live.' In the Middle Ages people would beat their breasts (where the heart resides) when confessing to such and lesser guilts, proclaiming: '*Mea culpa, mea maxima culpa!* (By my fault, by my great and terrible fault.)

As the heart is taken hold of, the depression reaches its culmination; it now also appeals more to the heart of another person. (In the case of mania increased involvement of the heart shows itself in the intensification that leads to frenzied rages.) A new source of suicidal thoughts opens up for depressive patients, and these may be particularly dangerous. The moral dilemma from which they arise once again points to the heart as the organ of conscience.

'My conscience smote me,' as the saying goes. The voice of conscience comes from the heart, and through it we hear the voice of the higher ego, which influences our lives through the heart. Patients suffering from endogenous depression often have hypersensitive consciences. They now hear only the voice of their earthly ego through the heart; cut off from the spirit, this ego feels itself to be shackled to the body through the liver and able to respond only to the painful experiences that arise through the pathological state of the soul.

Here the image of Prometheus bound arises; day by day the eagle of Zeus would eat of his liver, which then grew whole again during the night. Prometheus in his Titan pride had brought fire to mankind from the sun and as a punishment had been shackled to the rocks of the Caucasus. In melancholia or depression human beings feel themselves shackled to the body and the earth. Every day brings new pain; every morning no longer holds the glow of creative activity but is overcast with the darkness of night. The sun of the ego barely rises above the horizon of earthly existence, and the shadows of past deeds seem to grow infinitely vast with the light coming in at such a low angle.

The patient whose history is given above was a warmhearted person full of empathy whose childhood had been immersed in gentle warmth of feeling. She had chosen a family for herself where mind and heart set the tone. This kind of disposition and environment is characteristic in the history of endogenous depression.

There had been depressive moods, and the first real depression developed when she was eighteen; like those earlier moods it was the immediate reaction to something that had happened, in this case the death of her dear father. She missed her father for a long time, for he had been an indispensable stay and prop in her life. However, she did not lose her hold and the birth of her ego was not noticeably impeded. A sentient soul characterized by warm feelings developed under the impulse to help her mother. Yet there remained an inner need for the ego to lean on someone, and this led the patient to marry an older man. He was unable to take her father's place, however, for this was not in his nature, nor would it have been in accord with the patient's own inner development.

Initially the whole of her fully developed sentient soul, with the mind soul element already apparent within it, was totally given up to her children. She wanted to give them the same kind of home life as she herself had enjoyed. The crisis came when she was in her late twenties and early thirties and the children no longer needed her so much, so that the ego no longer had that support from outside. Neurotic heart symptoms developed as the powers of maternal love that were no longer needed acted back and put pressure on the heart. The intellectual and mind soul, which needs an ego that has grown strong in itself for its development, could not develop fully. Depression developed as a disease of mind and soul, and the neurotic changes in the heart that were the preliminary stage progressively developed into the heart symptoms at the level of body and soul of her depression.

There is however also another, more profound element to this;

without it, one might well imagine that cardiac neurasthenia would have been the only condition to arise. At the age of twenty-four, the patient had had a jaundice due to hepatitis, and this had not been completely cured. (Apart from the digestive problems this was also evident from physical investigations.) This liver disease is primarily ascribable to the unassimilated shock and grief of her father's death when she was eighteen. The fact that the depression developed three years after the nodal point of the twenty-first year in itself suggests that one should consider what happened three years prior to that nodal point. In this particular case the answer was that the event that had happened when she was eighteen, the organic 'shock to the system' she then experienced, was reflected via the nodal point to cause the organic illness three years after that nodal point. The psychological condition that arose from the organic illness was subject to the same law, for it emerged fully when she was thirty-two, i.e. four years after the nodal point of her twenty-eighth year, whilst she had had her hepatitis four years before that (see the chapter on *Ether Body and Metamorphosis*). Events are not always mirrored so clearly, and one usually has to think in terms of variations around a figure that serves as a guide for this mirroring of events.

In the case of this patient, a life crisis and the endogenous pathological process thus combined in their effects. The life crisis made the inner disposition of the patient manifest; the biography of the disease combined with the biography of the individual. A disposition to develop inwardness, to develop the intellectual or mind soul, was there, but the process was caught up in the tendency to take things the hard way, in a weakness in the assimilating function of the ego. That was the situation of the soul which arose out of the body when the patient developed her depression, a situation that seemed to overcome her completely.

However, endogenous depressions normally pass–at least in patients of this age group–after some weeks, months, and rarely years. Afterwards patients experience a new state of health that has greater value; this is often reported. They have come through, the life of the soul has been deepened, and these individuals now show greater inwardness and at the same time also are more active than they were before the illness. The liver and gallbladder system has become the instrument for ego and soul. The seeds they had within them have germinated: Taking thinks the hard way has become a way of taking things at a deeper level, or at least beginning to do so, thanks to the illness.

The meaning of depression in the context of biography reveals itself. This was also perceived by the psychiatrist W. Schulte.[60]

Different Forms of Depression and the Meaning of Depression
Melancholia, sometimes alternating with high spirits, is part of the biography of young, adult and ageing people. Different forms of depression develop in the course of life, and the most important of these are given below. Initially there are two polar opposites: reactive depression and endogenous depression. In the former, the liver, being the instrument of mind and heart, is in resonance with the soul's experiences; in the latter, soul life has to be in resonance with the pathological principles that rise from the liver into soul life.

Different combinations of these two trends produce different forms of depression, with one of the two poles always giving the dominant note. An endogenous component is thus diagnosed in reactive, and a reactive component in endogenous depressions. In the first case the liver is involved to a greater extent, as is evident for instance from the morning aggravation; in the second case events from the past play a greater role than in purely endogenous depression. The reactive mood changes and reactive depression shown by the above patient did have an endogenous component, and this is noted more and more frequently today. Even if the endogenous component does not show itself, involvement of the liver should be considered, as every reactive depression is a psychological illness; the involvement is however greater than with the simple resonance of low moods.

Neurotic depression is diagnosed if neurotic conflicts play a particular role, though there also tends to be an endogenous component and involvement of the liver. Exhaustion will of course also affect the central organ of life and of the life body so that exhaustion depression develops; the exhaustion causes congestion in the liver system. If the depression is masked by physical symptoms, we are dealing with masked depression, which is a common condition. Here the experience of being stuck, held captive in the body, is experienced more through physical symptoms; behind these, however, lies a depression, the treatment of which should be our main concern. Depressive states develop during the period of menstruation, for instance, and may become climacteric depression during the change of life. Involutional depression arises in conjunction with the physical regression that is part of the ageing process; the psychological aspect of this has been considered in the discussion of the problems of the mid-life period. In all these cases the liver is also involved, and this is of major significance in the treatment of the different forms of depression.

The incidence of depression in all its forms is constantly rising. About one per cent of the population suffer from endogenous depression, but the total number of depressions is many times that number. It has been stated that currently every tenth patient who goes to see a doctor (not a psychiatrist) suffers from depression and it is estimated that between 100 and 150 million more people in the world develop a clinically diagnosable form of depression each year.[67] The question arises as to the causes of this desperate increase. In the present context one would in the first place think of the human crisis that is in preparation when people are in their twenties and emerges when they are in their late twenties. The increase in the disease of mind and heart known as 'depression' indicates that this crisis is getting more severe and also points to increasing developmental disorders in the intellectual or mind soul. If these disorders are not overcome they may manifest as depression even in old age.

This leads to the final question. Are people quite generally taking things harder now when they reach this crisis? We have seen that taking things the hard way is one of the preconditions for the development of endogenous depression. The answer must be in the negative. The impression is that many people are taking things too lightly, and this superficiality is the principal danger in connection with the disorder in soul development.

The superficiality that arises as the sentient soul is held on to, does however lead to the same result as taking things the hard way, which is a tendency seen in the smaller group of 'liver' people who show greater depth of feeling. Superficial people who do not want to or are unable to assimilate what has happened to them, try to 'deal with them' rapidly at the surface. They will however only succeed in so far as the events leave the surface of soul life and sink to its depths. There the suppressed experiences weigh just as heavily, or even more heavily than things taken the hard way which the individual has not been able to assimilate. They are like food that lies all the more heavily on the stomach and the liver because it has not been chewed properly. The low mood that later rises from the injured liver into soul life will then impose itself all the more on the soul the more superficial and passive it has been in its development. The majority of depressions where the liver is appreciably involved no doubt arise by this route. Depression may thus be seen to have not only individual but also general human significance. Schizophrenia was shown to be the opposite side of the coin to the extreme intellectualism of our age. Depression, taking people to the depths, may be seen as the opposite side of the coin to superficiality of soul. It, too, may serve as a warning, and it may induce us to take the first steps to overcome superficiality and develop the intellectual or mind soul. This may help to prevent depressive illness for some people.

Therapeutic Aspects
The number of available physicians no longer matches the vast and steadily growing numbers of patients suffering from depressive illnesses, and this applies even more so to psychiatrists and psychotherapists. This means that in less severe cases every human being is called upon to help. In more severe cases a physician should be consulted if at all possible.

Before one tries to give help or assist in giving help, it is important to consider which type of depression we are dealing with. Initially distinction must be made between reactive and endogenous depression. If the depression is predominantly endogenous (the pure form is only rarely seen), medical treatment and remedial eurythmy based on the liver and the heart are the primary line of approach. Intensive measures of the kind described in the general section may be mandatory in really serious cases (mainly fever therapy and hydrotherapy).

Psychotropic drugs in form of antidepressants need not be used to the same extent as in schizophrenia. They should be reserved for serious cases where patients refuse food and there is a high suicide risk, and where the pressure under which patients feel themselves to be and the restlessness are so severe that the patients (and those around them) can no longer cope. In all other cases it is possible to go through the depression with the patient, using the treatments that have been mentioned as an aid. It has been found that these measures shorten the depressive phases and have a prophylactic effect concerning future episodes. If psychotropic drugs are given in high doses to suppress the depression, the gain we have spoken of above cannot be made through the illness. These drugs will also shorten the phases, but they may cause the depression to become chronic, having made it tolerable.

Art therapy should involve modelling exercises designed to let the plasticizing powers of the liver system to flow freely again. Sometimes painting is the better method for this, as this also directly activates the feelings. Music has a direct effect on the darkness that has fallen on the mind; as a rule one would start in a minor key and gradually change to the major. (The opposite with mania.) Speech exercises are clearly needed if one hears the low voice of a depressive patient that is lacking in energy; strengthening the voice one also stimulates will-based ego activity.

In the treatment of predominantly endogenous depressions great importance attaches to the attitude to the patient and the way the patient is addressed. Ignorance or impatience often lead to mistakes. Appeals to these patients to make an effort and pull themselves together are usually worse than useless. They merely encourage them to further self accusations. People who are seriously depressed really are unable to do what they want to do. The will is paralyzed just as much as an arm may be paralyzed at the physical level when the nerves have become badly inflamed. Would you tell such a patient to make an effort and would he then be able to move his arm? Quite the contrary. We would immobilize the arm until the inflammation goes down and then gradually start with movement exercises. In the same way patients suffering from severe depression must first of all be immobilized, perhaps even put to bed, and given treatment. Once they are better they will be asked to perform small tasks and offered a book to read (a serious, not a funny one), and one will note how they react to this. There is no point in giving orders, for patients have to learn to get their own will powers going again.

In serious cases treatment should always be under the guidance of a physician. Sometimes, and particularly if there is a real risk of suicide, patients have to be admitted to the closed wards of a psychiatric unit where they can be protected from themselves and given more intensive therapy. (In the case of the above patient this was not necessary.) In less severe cases outpatient treatment is the better choice, and efforts should be made to keep the patient at work if at all possible.

The key factor is authority wielded with love. The state of the patient, who has become more or less helpless and is profoundly depressed, is a partial regression to the situation of a child in the second seven year period who cannot manage on his own. Children of that age, when the foundations are laid for the development of the mind soul, inwardly long for authority wielded with love just as these patients do.

This kind of authority must also determine the approach to seriously ill patients. It is impossible to talk these patients out of their self accusations and guilt feelings, just as it is impossible to talk schizophrenics out of their delusions. It is more effective to assure these patients over and over again, in a loving but authoritative manner, that this phase will pass just as surely as the sun follows its course in the sky. The kind of image we have spoken of before, of the sun which may be obscured for us on earth but is never touched by the darkness, has also proved helpful. The kind of images used for educational purposes in the second seven-year period can generally be used to good effect with depressive patients even if the condition is severe.

If the depression is predominantly reactive, having arisen in response to a sad experience, the essential approach is to address the soul. One would try and help these patients to assimilate the painful experience. It is important first of all to take the time to listen quietly. The opportunity to describe the experience and talk about it in itself gives relief and improvement. This also applies to talking about suicidal thoughts, and patients suffering from relatively severe depressions should always be asked about these, using a gentle approach. Making use of therapeutic authority one can get patients who are not seriously in danger of committing suicide to promise that they will not harm themselves. That proved possible in the case of the above patient, though she had to be asked to come to the surgery quite frequently for this purpose. Patients are able to hold to such a promise to some extent. It has to be realized, however, that the risk of suicide is greatest when the will is still able to stir to some extent, or when it begins to stir again as improvement sets in. There are situations where no promise will help.

Sometimes it is also possible to look with the patient for the personal meaning of the depression once understanding has been gained. The question then arises as to what lessons for the future can be learned from the depression. What can I do to prevent it arising again–should my attitude to my partner change–should my attitude to myself change?

The first preventive measure would be to learn to assimilate better, and a good method is the reflective review of the day's events in the evening. Individuals who are disposed in that way will feel a desire to 'understand' everything. In severe depressions, however, when patients are incapable of any real sentient response, it would be best to start with a review of events in the evening. (Was there anything in the course of the day that gave me at least a little pleasure?)

A depression is always a challenge not only for the patient but also for anyone who wishes to help. It may also prove to have meaning for that other person. Anyone who tries to help someone suffering from depression may find that they themselves come to experience things in a more inward way and achieve further development. This should not, of course, be the aim when one desires to help. If we make a conscious effort to penetrate to the essential nature of the other individual we also receive impulses towards the development of our own awareness soul. As we make the effort on behalf of the other person our own awareness soul connects more with the heart and experiences itself at the centre of a great healing process.

The diagram shows that obsessive-compulsive illness has its roots in the mental image-forming process in the brain and takes place in the region of the lung, so that the solid element predominates.

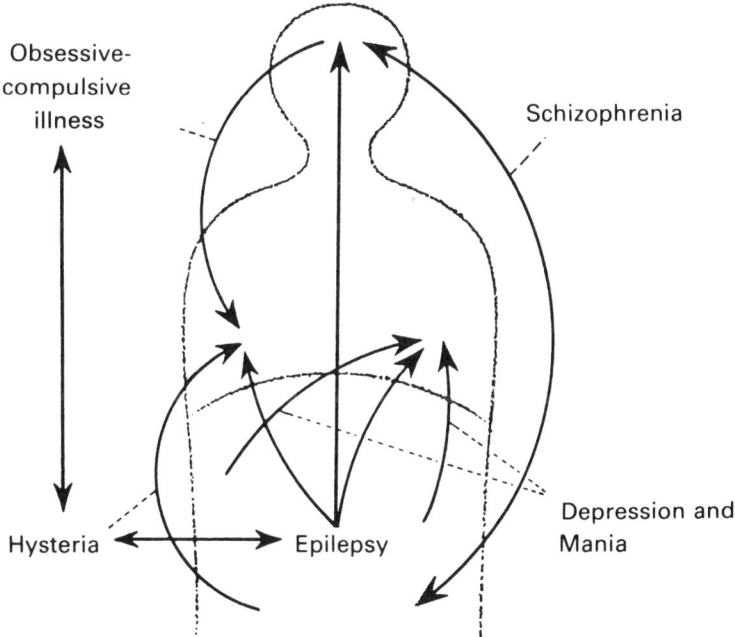

Obsessive-
compulsive
illness

Schizophrenia

Hysteria ←→ Epilepsy

Depression and
Mania

The keynotes for the dynamics of this condition are hardening and spasticity.

In the biography the illness characteristically has its onset between the ages of seven and twenty-eight, with a childhood form in the first seven-year period, which is also the time when the disposition to the illness, which only manifests later on in life, may be established.

Hysteria rises to the lung region from the instinctual life of the reproductive organs, and the airy element predominates. The keynotes for the dynamics of hysteria are dissolution and streaming out.

A vertical polarity to obsessive-compulsive illness shows itself, a polarity between above and below.

In the biography hysteria characteristically has its onset between the ages of fourteen and twenty-eight, with a childhood form in the second seven-year period; sometimes the disposition develops in the first seven-year period.

Epilepsy bases itself on the kidney and liver system, with failure to penetrate the airy and watery elements in those organs. Other organic systems may also be involved. The brain always plays a role (see the arrows going out from the word 'epilepsy').

The keynotes for the dynamics of epilepsy are stagnation and explosion.

Projected on to the plane, a horizontal polarity to hysteria shows itself, a polarity between inside and outside.

In the biography, epilepsy characteristically develops in the second seven-year period, the disposition having become established in the first seven-year period; childhood forms are seen between the ages of one and fourteen and also in the third seven-year period.

Schizophrenia develops out of a kidney system that is in polar opposition to the brain. The airy element with its dissolving properties is in opposition to the solid element. In this illness polar opposition is most marked and leads to chaos.

The keynotes for the dynamics of schizophrenia are schism and interpenetration.

In the biography, onset characteristically occurs between the ages of fourteen and twenty-eight, the disposition having been established between the first and the fourteenth year.

Depression and mania base themselves on the water organism in the liver and gallbladder system. If there is a swing to the solid element, the foundations for depression are laid, whilst a swing to the fiery element provides the basis for manic states. The heart is particularly drawn into the process which comes to its culmination in this organ.

The keynotes for the dynamics of depressive and manic states are stagnation and discharge. They go in the same direction as the keynotes for epilepsy, where the more powerful term 'explosion' has been used to indicate a more massive process.

In the biography, onset is characteristically from the twenty-first year onwards and into old age, with both the disposition and childhood forms developing between the first and fourteenth year.

The relationship between the clinical pictures and specific organic systems should be taken to be a dynamic one, and the clinical pictures should not be taken to be sharply defined. The organs and the characteristic psychotic disorders connected with them represent foci, with psychotic disease processes radiating from them, interpenetrating and mingling. It is essential to grasp the essential nature of the syndromes most immediately based on the organs so that the different components can be recognized and effectively treated.

The four organic systems that have been outlined above–other systems may be related to them–provided the basis for Husemann's most important work *The Anthroposophical Approach to Medicine*. These systems also proved fundamental in the development of an extended psychiatry based on the indications given by Steiner. (This is discussed in the chapter on psychiatry in the above work.) A psychology of organs has developed out of this.[1, 68] The literature on neurosis research sometimes also refers to four basic types: depressive, schizoid, compulsive and hysterical neurosis.[69]

It should always be remembered, when working with the characteristic psychological and psychiatric syndromes that have been developed out of this, that the ego is always the higher authority and ranks above all characteristic phenomena. The ego caused the illness to develop; it gives it its individual character and has its own individual way of dealing with it. It is however part of the essential nature of psychotic disorders that the ego becomes more or less submerged in them. To be able to help the ego to come to terms with the disease we must give serious consideration to the typical aspects of psychotic processes.

Chapter Six
*Art Therapy and the Constituent Principles**

As the name implies, art therapy has a specific therapeutic purpose and does not merely mean keeping oneself occupied with artistic activities as a form of 'occupational therapy'. We have seen that any genuine therapy is inherent in the illness itself and every illness calls for its own specific treatment. This also applies to every form of art therapy. Like the illness, it has its roots in the human individual and is developed out of the whole person. Below, examples given earlier in this book will be taken up and a summary outline given of the different forms of art therapy and their origins in a comprehensive view of the essential nature of the human being.

Once the initial problems have been overcome, any form of art therapy proves a source of the same kind of new vigour and refreshment that a refreshing sleep can give. The world has more colour and form in the morning than it had the night before. Sometimes we even wake up with a colour in our mind, or a particular colour stands out in the environment in a way that is different from the night before. We may wake up with a tune, a sound, or perhaps just a minor or major mood as we know it in music. A familiar poem or a verse from a song comes up and suddenly appears in a new light, sounding new and different.

*This chapter represents the revised and extended essay originally included in the Festschrift published to celebrate the fiftieth anniversary of the Friedrich Husemann Clinic on 30 March 1980.

Creative individuals gain impulses and contents for their work as artists from the night. Thus the painter Raphael said that having gone to sleep with a painting of a madonna left unfinished he suddenly saw it finished and complete before his eyes as he emerged from sleep. Many of Schubert's melodies came to him in his sleep, and Bruckner found some of the themes for his symphonies. Goethe woke with poems in his mind, sometimes writing them down as they came to him during the night.

Whence do these things come? Certain feelings connected with sleep may give a hint. The sensation of swimming out to sea or floating away as we go to sleep and the sensation of flying sometimes experienced in dreams indicate that soul and ego leave the living body during sleep and return to the original source of all earthly life, to the womb of the cosmos.[1]

Steiner said that his researches had shown that the soul and ego were in the world of the stars while we were asleep, in the spheres of the cosmos the harmonies of which the astronomer Kepler has described. It is possible to say that apart from those original sounds the soul and the ego are also influenced by original forms, colours and words in those spheres. Individuals may take some of this with them into the day, usually without being aware of it, though they will feel refreshed and filled with new life by the lingering effects of these experiences. The time spent in the cosmos also provides the basis for our ability to relate to the individual arts and for the ability, the organ, that permits us to be active in them.[2] The mythological story of the Muses who come down to men from the heights of heaven and bestow happiness on them with a kiss reflects something of this cosmic event.

It is now also apparent why everybody really has some artistic gifts and is able to do creative work in art.[3] All of us are able to bring artistic experiences with us from sleep, some more and some less so, and these experiences then rise up inside us and may develop into abilities. By morning they have first of all united with the sphere of our soul where it is always night, that is, our subconscious soul life. It is however possible to wake from this sleep as well, and we must come awake here if we want to appreciate art or produce works of art. This waking-up process does not occur of its own accord. It is a slow, gradual process and has to be brought about by ourselves in that we seek a relationship to art and become active in this field. If no such effort is made it might seem that we have no artistic gifts.

This also provides the answer to the obvious question as to why art therapy is important, seeing that the soul receives art into itself during the night. First of all it must be said that even our receptiveness is now at risk. Individuals who during the day are totally caught up in all kinds of concerns will be 'deaf and blind' to artistic impressions from the cosmos during the night.[2] To be ready to receive the 'gifts' of the night is is therefore important to practise the arts during the day.   ·

Every individual is offered gifts during life, and particularly during the nights, but we have to prove ourselves worthy to receive them. What is more, we must make something of the gifts we are given, make them our own, so that they will not be lost. Goethe's words, in his *Faust*, 'Whatever you inherit from your fathers must be acquired anew before it is truly yours,' may be rewritten with reference to the arts to say: 'Whatever gifts you received in your nights must be acquired anew before they can be yours.'

The need for this was less urgent in the past when gifts of genius that would remain for life were more common. Today experiences we have been given fade away much more easily than before, and 'gifts' of genius are lost before we reach the end of the twenties. Such gifts are most likely to come in childhood today, for children sleep more and thus maintain the link with the cosmos. Children are still allowed to live on things they are given, they draw 'as if in their sleep', sometimes showing marked genius. Adults have to practise hard to acquire such skills.

Different artistic impressions are received into different spheres of human nature during sleep, to rise again, or be gathered up again, from those regions during the day. Having risen, those individual artistic experiences have a healing effect on the regions from which they came. Below, these beneficial effects will be described in more detail, partly basing myself on the experience of physicians and clinics working with anthroposophical medicine. Mention should above all be made of the definitive work of M. Hauschka.[4] A number of other authors are mentioned in the References, as their works also provide initial orientation. Guidelines for art therapy to treat psychotic disorders were developed by F. Husemann at the Wiesneck Sanatorium, and by Mrs W. Husemann specifically in the field of eurythmy. After the deaths of F. and W. Husemann, physicians and art therapists at the Friedrich Husemann Clinic continued to use those guidelines and develop them further.[5]

The Individual Arts

Architecture

Initial orientation with regard to primary architectural forms is provided by the three dimensions in space: above and below, right and left, front and back. Going to sleep we sometimes have the feeling that we are expanding in all directions. On waking the experience may be of contraction from all points in space, sometimes as if falling into the body. Our physical bodies are built in three-dimensional space, they are spatial. The experiences of space we have during the night–the most basic of these have been mentioned–therefore come primarily from the body.

It is from the physical body, the house we move into again every morning, that we derive a feeling for architecture. Architecture we see in the course of the day influences the whole of our physical body and is perceived with the whole physical body.[6] We even feel physically uncomfortable if the architectural proportions are not very good. We duck involuntarily if a balcony is positioned in such a way that it looks as if it might come down at any moment. This is something we may experience even if we do not stand beneath the balcony and know perfectly well that the architectural design is such that it will not come down. If we intensely experience a Greek column our whole body will rise up straight, and we may feel tempted to fly away from our bodies when we see the flying buttresses of a Gothic cathedral.

Impressions like these have a positive or negative influence on health. Intense experience of a Greek column may enhance the power of uprightness in both body and soul. The perception of architectural forms may in conjunction with other measures have a beneficial effect on postural abnormalities in soul and body, particularly if these affect mainly the 'column' in our bodies, the spinal column. Compared to the experience of columns in looking at a Greek temple, when we are made to stand upright on this earth, the experience of a Gothic cathedral liberates us from elements that have become all too earthly because our souls are caught up too deeply in the body and in everyday life.

It is evident from the above that the impressions made on us by architecture, by lectures on art and by art tours improve not only the mind but also the health of body and soul.

Sculpture and Modelling

Temples and cathedrals also have statues standing in them, and the basic forms of these make up the art of sculpture. Plato perceived those basic forms in the Platonic bodies that were named after him-cube, tetrahedron, octahedron, icosahedron and pentagonal dodecahedron. Modelling usually starts with these five forms at the Friedrich Husemann Clinic. Another basic form is the sphere; this may give rise to an egg-shape, various animal forms and finally the human face. Movement enters increasingly into form in the process. Modelling different forms that are part of the human body (the ear of the person sitting next to one) increases the relationship to one's own body, which is also stimulated by the modelling activity of the hands.

Every night the sculptor in us, the ether body, receives new form impressions. A feeling for the art of sculpture develops from these. Compared to the bony skeleton and the 'architecture' of the physical body, the ether body comes to expression above all in the muscles which create the 'sculpted' body form around the bones. A piece of sculpture that has not turned out well or a lopsided vase may cause the muscles of the beholder to go into spasm. When we see a good piece of sculpture we feel more comfortable in the sculpture of our own body.

Modelling and sculpture support the ether body in its activity of building and repairing the body. At the same time the soul, whose life is rooted in the ether body just as much as in the life of the physical body, receives new form impulses. Modelling is specifically indicated if there is a tendency to dissolution in body and soul. At the physical level such tendencies may lead to ulceration, for instance, at the soul level to chaos in the soul, to the soul flowing out into the environment in a pathological way, and finally to psychosis. The structuring activity of modelling counteracts such tendencies from the ether body, and this is particualrly important in art therapy for schizophrenia and hysteria. In the case of epilepsy a thorough structuring process helps to penetrate the obstacles in the organism more fully.

Pottery goes hand in hand with modelling and sculpture. Here one gets a particular relationship to the human organs. If organs have become too porous and their fine structure has been damaged, moulding all kinds of vessels–vases, bowls and pots–stimulates reconstruction of the organs. Firing the pots in the kiln will finally enhance the process and this again influences the life of the soul. Wood carving, an activity where one is always taking away the material that hides the form, is harder work; it takes hoid of the individual right down to the bones. Taking the anthroposophical view one sees a relationship to the way bones develop: First there is a 'rough cast', and then certain cells continue to remove bony substance until the final structure and form of the bone has been achieved. Wood carving should therefore have a positive effect also on bone formation; at the same time it helps individuals to have more 'bone' to their souls.

Plaited strands interweave in different patterns to create a vase in ceramics. Patterns could also be drawn. Again a relationship arises to the structures of the body, for there, too, different structures interrelate. In the soul, thoughts interweave, support and correct one another. The positive effects of plaiting work–also when weaving textiles–may be brought into play where thoughts grow confused or a single thought has undue dominance and becomes a 'fixed idea'.

Weaving creates forms and images in a system of threads. In the human organism the nervous system represents a structure of threads spun around all organs and in the final instance cells; the creative powers gain orientation for their activities from this system. Weaving thus has a special effect on the nervous system that mediates formative influences in the body, and this effect also causes structuring impulses to pass from body to soul.

Drawing is halfway to painting, but there is an emphasis on outline; the therapeutic effect on outflowing tendencies in body and soul thus relates drawing to modelling and sculpture. In black and white drawing, using the shading techniques developed by Steiner for educational purposes, the hand maintains an angle that goes from above right to below left. This corresponds to the 'ego' gesture in eurythmy, the classical form of which has the right arm raised and the left pointing down. Like this eurythmy gesture, black and white shading therefore addresses the ego which inwardly and outwardly gathers the powers of the human being and directs them to a goal. This process is emphasized by maintaining the angle in black and white shading. Among the fine arts this is the discipline that particularly strengthens the ego.

Painting
Working with flowing colours we draw most immediately on the colour and light experienced during the night and this has an effect

on all processes in the body that are in a living state of flow. Particular importance attaches to the action on the blood as the vehicle for sentient responses and feelings. Impressions received through the eyes are taken as far as the blood. There they are completely absorbed into the astral body which evolves its sentient life in the element of air within the fluid element, i.e. in the internal respiration between blood and body substance. The soul, which is addressed from there, lives wholly in colour.[7]

We have seen that it is important for the life of the soul that as many sensory impressions as possible are transformed into living sentient responses; this applies above all to impressions received through the eyes, but also to the other senses. Sentient responses of this kind develop particularly if impressions are later recalled. Inwardly we are all of us painters when we cherish colourful memories. Every time we recall something we are 'picturing' it, even if nothing new is created in the imagination and we merely let the remembered images assume colour.

The physical effects of muddy colours and of bad colour combinations and pictures have an undesirable effect even on eliminatory functions; they may actually cause nausea.[7] Pure colours and good colour compositions on the other hand stimulate the movement of the juices, the blood and internal respiration. The effect ranges from the relief of spasms to new constructive impulses arising from the blood. Our organs are subject to a continuous process of destruction and continuously 'coagulate' afresh out of the blood; painting thus acts to support the modelling activity through which the new form of the organ arises.

Working with flowing colours gives life to sentient responses that have grown cold and grey, a condition often seen in psychological illnesses where there is stagnation or a frozen state of soul (particularly in depressive states and epilepsy). The enlivening effect on sentient responsiveness also gives new life to the faculty of forming mental images. It counteracts the hardening processes that lead to compulsive ideas. The activity of creating images in painting is the counter process to the passivity imposed on patients who have compulsive ideas. Painting thus also creates the conditions under which individuals may receive new impressions and assimilate them, a process that is made difficult when the 'grey' boredom prevails which is one of the main indications for painting therapy, and in obsessive-compulsive states.

Wet-in-wet painting with water colours is the method of choice, progressing to let the form of a flower, a landscape or a human form arise. The therapeutic effect also depends on the colours that are used. Thus the colour red has an immediate stimulant effect via the blood, blue is a soothing colour that acts from the head to reduce excessive blood processes that come to expression in excitement and aggressivity.

Layer painting, with one layer of colour put on another, always waiting until the previous layer has dried, clearly has a restful element even as far as the technique is concerned. Compared to the wet-in-wet method, one is encouraged to stand back and look at the picture as it is developing. This puts greater emphasis on the neurosensory system with its perceptive and awareness functions.

Three soul principles are directly addressed during sleep by painting and by the arts that are discussed below–in contradistinction to architecture, sculpture and modelling. It is the sentient soul that receives primal impressions relating to painting whilst it is in the cosmos. On waking it shares these impressions with its darker brother, the sentient body, which remained behind in the body and now acts via the internal respiration to produce the effects that have been described.[2] During the day, the inward sentient life of the sentient soul shows a primary relationship to painting. Painting therapy regenerates the life blood of this soul principle and is therefore, together with remedial eurythmy, the method of choice for developmental disorders of the sentient soul (and sentient body).

Music
On waking in the morning, human beings bring with them musical experiences gained among the harmonies of the spheres, and these provide the basis for a feeling for music. All forms of music act primarily on the respiration of the lung; we experience this when we take in music and observe our breathing as we do so. Our breathing goes with the music and can be liberated and given wings by it. That, however, is merely the beginning of a more subtle process that starts with the respiration and continues on into the inner organism. (This is not the connection between respiration and the blood.) The respiratory activity of the lung is taken up in a different, i.e. rhythmical, form by the cerebrospinal fluid. As we breathe in, the fluid rises and so does the brain which is floating in it; as we breathe out, both go down again. The rising and falling tide of this 'respiration' is influenced even at its subtlest levels by music.

A connection thus arises with the image-forming faculty that is served by the brain. This has gained colour through painting and is now set in motion through music. The train of thought set in motion by the will may thus be given wings; the feelings that always accompany thinking activity are given new life. This is also where the therapeutic effect of hymnic music on obsessive-compulsive states comes in. A form-giving relationship is also established with the will itself and at the physical level with limbs and metabolism. Acting from the middle, this breathing process brings harmony to the circulation of the blood.

To be able to assess the specific effect on the feelings, it has to be realized that compared to all the other arts music achieves the greatest inwardness. Painting arouses sentient responses addressing the outside and the inner world; music takes us to a life of feeling that circles within itself and forms the basis for heart and mind qualities. Music lives primarily in the human heart and mind, and apart from the liver system, the inner breathing rhythm that swings to and fro within itself largely provides the physical basis for heart and mind qualities.

Keeping things alive in mind and heart always has an element of reflection to it. A rational, organizing element lives in mind and heart just as much as in music, where it comes to expression in the mathematical relationships that are present everywhere in music. Examples are the ratios of pure intervals and the seven notes in a scale. Even the beat brings an element of order into the flow of music and into the flow of the feelings and of will impulses that go with music.

Music brings clarification and resolution or clarification and firmness to mind and heart, depending on the way in which mind and heart are 'out of tune'. If individuals have developed pathologically low spirits, if they are suffering from 'depression', music in a minor key will help them to put the lowness at a distance. As the music gradually changes to the major key the mood is lightened. Individuals who are not depressed but in pathologically high spirits and suffer from manic states are helped to make the mood objective by music in a major key. A gradual change to the minor key helps them to find themselves. This also applies to the outflowing one sees with hysteria. The torn state of mind that is so marked in schizophrenia is made objective through dissonance. Dissonance changing to consonance brings further healing to a soul that is torn in itself. However, music rich in dissonance that is not given proper artistic form and badly played music cause disruption all the way to the head and result in headaches. Therapeutic music on the other hand has a health-giving effect not only on breathing but also on the function of the brain, and this is easy to understand if one considers the way the breathing rhythm continues into the brain.

An instrument that gives special therapeutic qualities to music is the lyre. The sound of a lyre has more inward than outward vibration and lives more in the breathing of the spinal fluid than any other kind of sound. That is the region where we all have a lyre within us: the nerves arise in pairs all along the spine and extend from there into the organism, and it is not by chance that that is the image of a lyre that has been opened out. The spinal fluid passes along these nerves as it rises and falls, and 'the nerve is set vibrating' just as strings of a lyre are by the fingers passing across them.[8] Something of this comes to awareness when we hear music and feel something of a subtle thrill pass through the body.

It has already been said that the nerves have not only perceptive but also form-giving functions. Music therefore not only lifts our spirits but has a formative influence via the nerves that passes via the spinal marrow to the whole body. Specific music therapy is therefore also indicated in physical disorders where form principles are affected. Basing itself on the breathing of the spinal fluid, the melody as the thought element in music relates more to the head, whilst the rhythm in which the music moves relates more to the limbs. Melody and rhythm are united by harmony, and this in turn offers support to the human mind and heart in its search for harmony. The harmonizing effect of music on the circulation can be understood on this basis. It is a dynamic rather than static harmony of heart and mind that is always arising anew. Because of this, music also relates to the physical heart.

The term Steiner chose for the 'musical' principle in the human soul: intellectual or mind soul, points to the union of intellect and mind. Music will encourage this soul principle to move reflectively within itself. The intellectual or mind soul also presents the soul space where music sounds and from where it takes therapeutic effect.[9] Dwelling in the cosmos during sleep, this soul principle absorbs the harmonies of the spheres that may then echo through it during the day.[2] Music therapy and remedial eurythmy are particularly helpful with developmental disorders of the intellectual or mind soul. Music thus gains special significance for the standstill in soul development that initially affects the intellectual or mind soul.

Speech and Poetry

'Words from the world of the spirit', original words that human beings are able to receive during the night, awaken a feeling for speech and poetry. Music gave the greatest degree of inwardness; now the movement is to the outside again. The word has to be 'carried through' from the innermost soul and body into 'day consciousness' and the world.[2] New awareness arises through contact with the world, a new will exerts itself in the world. We recognize this from the mental images that language conveys and from the will activity that gives impulse to speech.

It needs feeling, however, to establish a direct connection with another person and the world. This connection is more powerfully effected out of the centre of the human being, out of the ego, if language and speech emphasize their origin from the heart. The heart imposes rhythm on our breathing through the rhythmical pulse beat. 'Human beings ... infuse their words and sentences with the quality that wells from the heart.'[10] The central expression of this is the physical warmth and the warmth of soul in our speech.

When we speak, words are formed in the stream of exhaled air. The inner warmth of our hearts also goes out to the world in our words. The lung exhales not only carbon dioxide but also body heat that has reached the lung with blood rich in carbon dioxide coming from the heart. The ego can allow itself to be borne on this heat in the act of speech, and out of the heart transform it into warmth of soul. Steiner used to speak with a conviction coming from warmth of the heart that was most impressive.

Words spoken like that, coming from the heart and going to the heart, avoid the predominance of the will pole that shows itself in an emotional way of speaking, and also the one-sidedly intellectual way of speaking that comes from the head. Speech may be experienced as being too thick when it is emotional, and too thin when it is intellectual. Hollow emotional phrases combine both one-sided forms without creating the balance between them. It is laid on thick but to the inward ear it sounds thin; the healthy feeling of the middle sphere cannot keep up with this and is hollowed out. Creative speech must find the right way between the two extremes, filling abstract images with warmth, and using warmth to bring soul qualities to instinctive will impulses and calm them down.

Speech really always wants to become dialogue with another I, another person, who may respond in silence merely by listening. Reciting poetry should also always lead to a silent dialogue with the audience. Reading poetry on one's own should turn into a dialogue with the poet. Dialogue, this time organic by nature, develops even at the embryonic stage. Recent researches have shown that the child hears the mother's heart sounds in the womb and in a way responds to them. This response consists in developing the organs of speech, and particularly the larynx which later creates speech on the stream of exhaled air. The muscles of the larynx correspond to the internal structure of the heart muscle,[11] so that the connection between speech and the heart shows itself even at the organic level.

Creative speech, an art form developed by Steiner, also provides an effective form of therapy. The therapeutic action originates in the middle region and influences the physical and psychological middle via the breath.[12] At the centre of this therapy is the action on the sphere of the heart and on the ego in its activity in the heart. Arising from the heart–speech therapy also has a positive effect on the physical functions of this organ–the action of speech therapy extends up into the head and down into the metabolism through the breathing process. From the heart a way is then sought into the world. Speech exercises can give immediate help when our thinking is turned in on itself too much, the will is inhibited or too powerful, feelings are anxious, shrinking back from the world, or flow out into the world. Apart from brain disorders causing speech disorders that respond to speech therapy, soul conditions can also cause speech disorders that call for speech therapy. They may not cause major speech defects but undoubtedly always do so in subtle ways. Speech has a healing influence that arises from the middle, the ego, in such cases and permits a new relationship to be established with the world.

Listening to poetry and speaking it themselves, patients also learn to open up the ego to spiritual contents again. Subjective problems can be objectively presented in artistic form. Lyrical poetry comes from the innermost soul; it helps patients to reveal their innermost soul when this has been shut away from the world too much. The images presented in epic poetry take individuals farther out into the world and relate that world to their inner life, encouraging the memory functions performed by the ego and bringing order into them. In dramatic verse, individuals address themselves to others in translated form and learn to speak to others again if they have lost the ability to do so. They learn to express themselves in the context of the given role.

Speech and poetry are the arts that above all address the ego out of the soul; the ego speaks in the person who is speaking. They also address the conscious awareness that lives in the ego-centered soul. The awareness soul is the immediate soul organ for speech and poetry; creative speech and poetry are the arts that help its development in the most immediate way. Developmental disorders of the awareness soul are thus an indication for speech therapy as well as remedial eurythmy.

Singing Therapy
The sphere of action of this therapy lies between speech and music. When words are sung, the healing process enters more deeply into mind and heart and is then released into the world by the word out of the movement in a closed circle that represents musical experience. Warmth of heart in both body and soul again plays a major role. Singing makes it easier for people to give vent to everything that fills their heart and inmost soul. Some patients therefore find it easier to learn to sing first and then use creative speech. Singing may create the transition from music therapy to speech therapy.

Eurythmy
If we ask ourselves which kind of night-time experiences arise again through eurythmy, we come to see the following. Initially we perceive an architectonic element in the orientation in space given through eurythmy, for instance in different positions and groupings. Forms relating to sculpture and modelling arise when a gesture creates a speech sound or a musical sound in the air. Steiner, who created eurythmy, also called it 'moving sculpture'. The colours we use in painting here show themselves in the garments and veils and in stage lighting; they are inwardly alive in anyone doing eurythmy in the colour sensation that should go with every exercise. Musical sounds are made visible in one form of eurythmy, the sounds of speech in another. Every sound is represented by a particular movement, bringing to full development the processes that are only suggested by the movements of the larynx and other organs of speech when we speak or sing. Eurythmy gives expression to the invisible gestures of song and speech that lie behind the hints at movement in the organs of speech.

It is evident from the above that eurythmy combines all the arts; to some extent this also happens with other forms of art (e.g. there is a musical element in painting, and tone colour or timbre in music). What is more, the human organism, which started to be the instrument of artistic activity in speech and singing, entirely becomes the instrument in eurythmy. Every gesture extends not only into physical space but into the spheres of all the arts, to establish a new link with the eurythmical consonance of the arts that was experienced in sleep.

Watching a eurythmy performance also makes special demands on the audience. Mere onlooker awareness causes us to miss the most important element. We should actively enter into every eurythmy gesture and slip right into it, so that we are involved in the way it arises. To some extent we are asked to identify ourselves with any art form that we enjoy, but in the case of eurythmy this identification is particularly important. If we keep making the effort to enter into the eurythmy we see performed we penetrate through the image that presents itself to the movement itself and experience something of what eurythmy is able to bring out of the cosmos.

Eurythmy is thus able to bring all the arts to new life, and a bond exists through this with the earliest art performed by human beings, the temple dance, which also encompassed all the arts. Within the architectural form of the temple with its statuary, the gestures of the dancers in their coloured garments unfolded, accompanied by word and sound. People then felt that this art had been created by gods and was given form and design by their priests; individual experience and active design were less important. Eurythmy creates a new link with the cosmos for human beings, a link that is just as comprehensive as that which existed in the ancient art of the temple dance. Now, however, the essential individuality of human beings experiences and creates the things we have received into ourselves from the source springs of the arts.

Remedial Eurythmy
The art of eurythmy becomes therapy in remedial eurythmy.[13] The transformation of art into therapy is more obvious in this case than with the other forms of art therapy we have been considering. Steiner asked for strict distinction to be made between eurythmy as an art and remedial eurythmy; the two should never be mixed. Remedial eurythmy therefore is not a form of art therapy in the narrower sense. Movements performed in remedial eurythmy lie between art therapy

and medical treatment, for they concentrate on a particular disease process or diseased organ to a much greater extent than exercises based on the other arts do. It therefore acts very much like a medicament, but from the limbs rather than metabolism. This being the case, remedial eurythmy should always be prescribed and monitored by a physician. Remedial eurythmy nevertheless always has its roots in eurythmy as an art, and remedial eurythmists therefore need to be fully trained eurythmists before they receive specialist training in remedial eurythmy.

Eurythmy as a whole gives health to human beings. In remedial eurythmy, specific speech and sound gestures are used to concentrate this effect on organs and organic diseases that have an essential relationship to them, similar to the relationship that exists to natural substances of which medicinal use is made. Each sound gesture reflects something of the original image of the healthy organ that has once been created by the creative Word, the Logos as it is called in the first chapter of the Gospel of St John. Acting on the diseased organ via the limbs, the sound-gesture helps the organ to recall the image in which it was created and be restored to health. Apart from improved well-being, this shows itself in the way the gesture which the diseased organ caused to be pathological is restored to health, and this is the point where remedial eurythmy is initially brought to bear.

The original image and movement of an organ come to expression in its form and function. The gesture of the consonant 'B', closing in around the body, reflects and supports the form principle of a healthy kidney when there is a weakness in this organ (e.g. in a case of nephroptosis or downward displacement of a kidney). At the same time this gesture enters into the movement that exists in the process of elimination where material that has been eliminated from metabolism is concentrated from the periphery of the body towards the kidneys, where urine is formed. The 'B' will therefore also act on kidney function and noticeably stimulate the production and excretion of urine.

People may become so excited that they are beside themselves, and this is due to the kidney. The 'B' gesture helps the concentration of soul powers as much as of urine; apart from excretion there is enhanced 'incretion' of the astral body, i.e. the astral body is assisted in returning to the protection of the body when an exciting or shocking event has caused it to leave that protection too far behind. The effect of this gesture extends even to the condition of being beside oneself seen in schizophrenic psychosis.

Compared to the action of consonants, vowels directly address the soul in its experiences. People who are beside themselves with excitement not only lose contact with their bodies but their souls also get out of touch with the world. If this continues for some time, they may get caught up in themselves in spasm at the level of both body and soul. In these and many other cases the vowel 'A' (as in father) helps the soul to open up to the world again by penetrating the body, its mediator, in a healthy way. We experience the initial stage of this when we hear or speak an 'A'. It is not for nothing that 'A' is the vowel of amazement and marvel and marks the first step in the soul's relation to the world. This experience is heightened in the eurythmic 'A' gesture, with the arms extended and opening out. A specific form of this done as an exercise concentrates the experience in the kidney system and its problems at the level of body and soul. In general terms, the experience of radiance received that is gained through 'A' lets light enter again into a soul where the darkness of disease has fallen. The gestures relating to other sounds have similar effects on other organic systems and functions that have become diseased due to psychological disorders.

Eurythmy is taken into the ego-filled soul and reaches the physical body via the ether body. It unfolds as a mobile three-dimensional form, like moving sculpture, in the sphere of the ether body which has a profound relationship to all forms of modelling and sculpting, as we have seen. This is the sphere where all the constructive powers that mediate between body and soul have their source, and here eurythmy unites the other arts that are individually addressed by the different forms of art therapy.

Gymnastics
After the body of creative powers, the physical body, too, may be used as a therapeutic tool. Moving sculpture is followed by the 'moving architecture' of gymnastics which may thus also be regarded as a form of art therapy. Bothmer gymnastics, a method originally developed in Waldorf education, have proved effective and have been used in the treatment of psychological disorders for more than twenty years at the Friedrich Husemann Clinic in Germany.[14] Using this method, human beings themselves produce architectural forms that are taken up by the physical body and influence the physical body. The gymnastic movements are related to the three dimensions in space

and bring these and their metamorphoses to experience. Having re-established a healthy relationship to the physical body–particularly through remedial eurythmy– patients doing Bothmer gymnastics experience a new relationship to the outside world in terms of space through their physical bodies.

Prescribing Art Therapy

The way in which the arts have been related to the essential principles of the human being in the above may provide the basis for prescribing. It is not the only way shown by Steiner but certainly the most differentiated one. It is important, however, not to become rigid in relating particular arts to particular principles. Every art form influences the whole human being, though it initially addresses itself to a particular sphere. Nor does a particular art form act only on essential principles or soul elements that have already come to birth. Work done in the sphere of the arts encourages not only underdeveloped principles and elements but also those that are as yet been unborn. Taking account of the relationships established in the above, it thus also has relevance for children and young people.

In case of illness, art therapy should always be prescribed and monitored by a physician. The wrong kind of art therapy is not merely useless but may actually prove harmful, so that it is necessary to work to the specific indications which physicians are able to gain from an overall view of signs and symptoms in body, soul and spirit. The outline that has been given above provides pointers for those specific indications that must then be given more depth and relevance by physician and therapist working in collaboration.

Patients with outflowing souls mainly need sculpture and modelling, for example, whilst patients who have developed an inner rigidity or spasticity mainly need to paint. When people suffering from acute psychotic states are quite beside themselves it will first of all be necessary to lay the basis for renewed incarnation; after this, they may gain a healthy relationship to the world again through speech therapy. In the same way gymnastics may be used to consolidate the incarnation process once this has been initiated by remedial eurythmy. As already stated, singing is usually indicated before speech therapy starts. These few examples may serve to show how differentiated indications for art therapy can be developed.

If only one form of art therapy is available, we would try and have it done in a way that meets the indication in the individual case. Outflowing patients would thus be given painting therapy with a marked form element; with patients held in spasm, the emphasis would be on flowing forms in modelling; music and speech therapy may stress the plastic or the painterly aspect, depending on what is needed, to mention just two of the many aspects.

Looking Back and Looking Ahead

We have reached the end of a particular route we have taken. Architecture took us into the world of three-dimensional space that is also part of our physical bodies. Sculpture, modelling and related arts made us enter into the organism of creative powers in the body that acts in conjunction with nature. In painting we lived in a relationship to the outside world that arose from inside us; music took us into an inner experience that was a movement closed in on itself. Speech and poetry established new connections with the outside world, this time out of ego powers. With eurythmy, this world-related ego then entered into our inner life and penetrated as far as the organism of creative powers again, finally to return to the physical body in a new way through gymnastics.

Having followed this route we find that the initial thesis does indeed hold true. Exercises in art are not something foreign that is brought to human beings from outside; instead it corresponds to man's essential nature, and through the exercises this essential nature gains new sustenance from the very sources of its existence. These sources open up when we are asleep at night. They pour their waters into the subconscious regions of body and soul from where they may rise again in artistic form during the day. Connecting ourselves to the sources by being receptive and doing the necessary exercises we also unite with our cosmic reality.

This relates not only to our present night times, however, but also to the time when earth and humanity were much more part of the whole cosmos than they are today. The story of Paradise, of the Golden Age, reflects something of the earliest childhood of the human race that is partly repeated in the childhood of every individual. Then there was no illness and no sin. The arts have preserved something of this for us.

When illness has taken us too far away from the life of the cosmos and from nature, artistic endeavour brings back an echo of the original health and innocence that existed in Paradise. When illness makes it impossible for the soul to free itself of the body, or when we are beside ourselves and in danger of losing the connection with the body, art therapy helps us to restore a healthy connection. The therapy repeats in a specific way the healing process we undergo every morning, a process arising from the sources of the night and the spheres of the different arts. In prescribing art therapy we are addressing the healthy individual inside the sick individual and encouraging self-healing processes that work in conjunction with the cosmos to bring healing.

All this can only be achieved, and we will only be able to draw on the resources of the cosmos, by patient and persevering practice. We know that in the twentieth century adult human beings essentially no longer receive gifts from heaven. They are challenged by the ego to unite themselves with the spirit of the cosmos and the spheres of the arts within it in ever new ways. This applies to patients and also to the art therapists themselves. Unless one continues to take in art and be creative in art over and over again, one cannot give art therapy to others, for art would then degenerate into mere technique. Having drawn on the resources of art, artists working as therapists then have to sacrifice their own creative activity and concentrate completely on their patients to help them to recovery. To relate properly to the disease in question and its treatment, it will also be necessary to work in close collaboration with the patient's physician.

It is not the aim of art therapy to produce works of art. The emphasis is on practising the art in question, for that is the healing element; if possible a certain rhythm should be introduced. Carefully chosen exercises in a particular art open up a source of new life for the sick individual.

After the first few attempts, individuals practising an art will also come to experience the sheer beauty of art. Whenever we perceive beauty something shines through where we are aware that this is more than I am, it follows certain laws and yet is full of life, and it is an integral whole. A glimpse of divine life is gained, a touch of the living spirit. This does not necessarily mean that one's Muse will give inspiration. It needs only the whisper of her wings to give us an inkling. And that whisper comes with every exercise in the sphere of the arts. Every one of us can today experience something of the grace that blesses the efforts patients make in this sphere and finally lets them experience the blessing of being healed.

Chapter Seven
Soul Development and Spiritual Training

It has been shown that human biography essentially serves spiritual development, though this does not have the support of the world that was given to soul development. Dangers threatened soul development even in the twenties, and individuals then needed the spiritual impulses that the world provides. They had to take hold of those impulses of their own accord, however, and spiritual development, too, is only achieved by training oneself spiritually in a wider or a narrower sense. Examples have been given in the chapters on the treatment of developmental disorders.

Individuals who want to go on learning from life may enter the school of the spirit. Preparing themselves and opening up to everything that comes from the world of the spirit through their higher egos, people help themselves and then also the world. Spiritual development generally only shows itself to be distinct from soul development in the forties, but it can, and indeed should, take its beginning in the process of soul development.

Rudolf Steiner described the spiritual training to be gained through anthroposophy which will be considered here in relation to soul development. It is fundamentally different from the Eastern disciplines that are generally known today. It differs primarily in that it bases itself on the level of conscious awareness that has been reached by the peoples of the Western world today in that they have come to awareness through the world that is perceived through the senses. Training will enhance and extend this awareness for the individual. Essentially it does not convey spiritual scientific contents as a body of knowledge, but shows a way of perception and insight where the assimilation of these contents leads to the transformation of thinking and other soul faculties. This assimilation in itself is spiritual training and remains the only form of spiritual training for many people who find it very helpful for their development. Further enhancement comes with meditative exercises, and these may give direct access to spiritual contents.

The path of spiritual training becomes a biography of spiritual training that may be seen to relate to the earthly biography and to soul development. One aspect of these relationships has already been discussed, and it has been shown how spiritual training or its fruits can further development. This chapter will show how spiritual training also is furthered by the soul elements as they develop, and it will be seen that this is subject to certain laws.

Spiritual training in the anthroposophical sense characteristically tends to start only after the birth of the ego in the twenty-first year of life. Occasionally it will start in the years of preparation for that birth, i.e. in around the eighteenth year, but only rarely before that.

Sentient Soul and Spiritual Training

Once the contents have been taken up by mental powers capable of judgement, the sentient soul initiates the process in which they are absorbed in a living way. The first sentient response in which the sentient body opens up to the world is wonder and marvel. If individuals do not develop this, spiritual contents are not absorbed with the whole of soul life and may remain confined to the head together with the ego and the soul. Experiencing the magnitude of a spiritual reality and reflecting on this, we also develop a feeling of awe and reverence. This protects us from the intellectual arrogance where we consider ourselves better than others because we know so much. Such a conviction means that we are again limiting ourselves to the head and considering ourselves superior to others, thus blocking our own path to further development. Speaking of the stage where we wonder and marvel and feel reverent awe in spiritual training, Steiner said with reference to the arrogance that may arise instead that people then 'merely become sharp-witted'. Progress will be made if the capacity for wonder and reverent awe is fostered.[1]

The first stage in spiritual training thus consists in developing a differentiated sentient response to the contents of our perceptions. This kind of living experience actually takes us back to childhood.

'To gain intuition of non-physical realities, one needs to some extent the freshness and perkiness one had as a child'.[2] Is it not true to say that in about the twentieth year of life the newborn ego experiences a second childhood as it makes its first attempts at walking in a world that has now opened up to it?

Children take in everything as if for the first time. They assimilate subsequent impressions on the basis of those gained earlier and may indeed already have developed an organ for them, but anything new they take in is always something fresh and new. We may try to develop the same attitude to spiritual contents, taking them in as something completely new that needs to be assimilated anew on each occasion.

The fruit of such endeavours is the 'open-minded receptivity' that Steiner considered to be the first goal to be achieved through basic exercises in spiritual training. The starting point are the impressions gained in the sentient soul. 'Every draught of air, every leaf on a tree, the babbling of an infant' should tell us something new. This should then be weighed against past experience.[3] More developed soul qualities are already coming into play at this stage, and the living memory of the sentient soul provides the basis for these. The ego is now called upon to create a balance between the new and things learned in the past, and this is probably also a reason why this particular exercise is only the fifth in the sequence of preliminary exercises.

The intellect will sometimes discover that the spiritual contents we are taking in are contradictory, something that happens quite frequently when the path shown by Steiner is followed. This is not a bad thing and it should not be suppressed. We should let the contradiction stand and let the sentient soul guide us as we experience the new element that arises out of the contradictory situation. It is this experience that will later on lead to the discovery that this was indeed a new aspect of the object under consideration and that this could indeed present something new to us. If we get caught up in the contradiction, we are trapped in the head's attitude of antipathy, which of course has its own important role to play. The first steps towards perception of the living reality will come when the differentiating function of antipathy is brought together with the sympathy process of making things a new experience in a process that also involves the time element. Perceptive awareness based on thought is complemented by the immediate experience in the sentient soul that leads to perceptive understanding of the new aspect.

The intellect, which is part of the next soul element to develop, is thus involved in spiritual training from the beginning. Does this mean, though, that the spiritual content is understood? With everything we take in through the sentient soul we must learn not only to let contradictory elements stand, but to give up the desire to understand everything immediately. Steiner spoke of the importance of things that are not understood. This was with reference to education. Children are sometimes taught things they cannot and indeed should not yet understand; they will only have the maturity to understand them at a later age.[4] Such maturing processes apply not only to children and the childlike ego of the sentient soul as it receives new spiritual contents. Things that are not understood lie in the subconscious depths of a sentient soul that is alive like seeds that will later germinate and grow into plants to yield fruits we would never have expected from those seeds. If the seed were to be 'understood' and assimilated immediately, many things could never come to flower and to fruit in the process of soul development.[5]

There are, however, two preconditions. In the first place the content we have taken in without understanding it must be a living seed with germinal qualities and not an imitation made of stone, i.e. an idea that remains abstract. Secondly, the soil receiving the seed must be alive, too, and must continue to be cultivated when we are no longer young. Steiner often spoke of the sentient soul in these terms, even if he did not explicitly name it. Thus he wrote in his *Letters to Members of the Anthroposophical Society*, 'If we let the beauty and sublime nature of the physical world enter into our hearts and minds they become a source for sentient experience of the spirit.'[6] The words 'sentient experience of the spirit' point to further, spiritual development of the sentient soul.

A sentient soul that is becoming spiritual lets feelings enter into it that are filled with the spirit. One of these is the feeling of inspired enthusiasm where we can have immediate experience of the spirit entering into us. It is important that the words used reflect the true state of affairs, i.e. that feelings are truly filled with the spirit. We are not referring to the wild, uncritical enthusiasm that clouds our vision and easily arouses antipathy in others. Enthusiasm aroused as the sentient soul responds to the spirit is no transient ardour that flickers into life now and again; we may compare it to a flame that gives warmth and illumination and helps the inward growth and ripening of things that we have not understood.

Intellectual and Mind Soul and Spiritual Training

Using the re-flective powers of the intellect, spiritual contents conveyed to us should be seen in a clear light from the beginning and
taken as they are, before we proceed to assimilate them through sentient response and thought. Steiner repeatedly spoke of the need for
'sound common sense', saying that this would help anyone to understand the findings of spiritual research. 'We may say that everyone
can understand the insights into the world of the spirit that have
been gained when they are conveyed in the form of ideas.' It needed
'an open mind and sound common sense, though this must search
and explore the soul in some depth.'[7]

The words 'open mind' immediately give pause for thought. Is
there still such a thing as an open mind today? And 'sound'? Kuehlewind is perfectly justified in saying that sound common sense is even
rarer today than it was in Steiner's day.[8] He refers to statements
made by Steiner that indicate that even in those days there had to
be reservations when it came to sound common sense, and in the
final instance these led to the statement that brain-bound thinking
cannot understand the ideas conveyed out of spiritual science.[9]
Steiner came to the conclusion that 'such common sense has to be
acquired today, and this takes effort.'[10] It can be done, thanks to the
'new concepts' that spiritual science presents, but these must be
recreated by the recipient and acquired through meditation.[8]

The problem of the lack of sound common sense may also be considered in relation to the general human crisis experienced in the
late twenties, a crisis that threatens to inhibit development of the
intellectual or mind soul. We have seen that the ego must develop
new activity in order to disengage soul development from physical
development which at this stage goes towards consolidation and later
towards disintegration. This also applies to thought activity, for if
it does not disengage it will continue to be brain-bound and therefore
will no longer be sound (see the chapter on *Psychiatric and Nervous
Diseases of Old Age*). We are no longer merely given sound common
sense today. Accepting spiritual contents with a clear but active mind
that is supported by the sentient soul, we also create the necessary
precondition for the development of healthy common sense in the
intellectual soul, with the sentient soul also providing the basis for
open-mindedness.

The subsequent assimilation of those contents by the intellectual soul depends largely on the other aspect of this soul principle, the element of mind and soul. Steiner's words, that the intellect 'must search and explore the soul in some depth,' point to this. Any exploratory effort is met with an inner response from the element of mind and soul, though in many cases this, too, must first be acquired or at least deepened. Steiner again and again called up the mind and soul element in conjunction with anthroposophy. In the first of his Guiding Principles (or *Leading Thoughts*) he said: 'Anthroposophy can only be given due recognition by individuals who must look for the things they find in it out of their minds and hearts.'[11] From the very beginning the intellectual and mind soul also makes its 'mind' aspect come to expression through the receptive sentient soul.

Following the development of the ability to marvel and feel reverence, Steiner refers to the third preparatory stage as 'feeling oneself in wisdom-filled harmony with the laws of the universe'. The intellectual and mind soul lives in harmony between feeling and law. Out of this harmony it brings about the growing inwardness and inner mobility that are needed for the meditative assimilation of spiritual contents and for meditative life altogether. Any form of meditation is in the first place 'contemplative reflection'[11a] in the intellectual and mind soul. The mind soul creates the inner space for meditation; its intellect watches at the gate (or ought to) so that no unhealthy mystical element enters the inner sanctum.

If this does happen, the danger that arises when individuals want to linger in the sentient soul rather than progress in their spiritual training will persist. The life of a sentient soul that is held fast in this way may degenerate into an intoxicating rapturous state and if the intellectual and mind soul is not very strongly developed this may assume mystical characteristics, in which the individual shuts himself off from the world. Another danger lies in the drying-up of sentient and feeling life in the light of awareness that comes with spiritual training–we have already spoken of this. The first danger arises in the sphere of emotions and drives; the second again has its origin in the head. Both show how important it is to create a protective space inside, a ripening space, and how important the 'calyx' stage of the intellectual and mind soul is for the 'flowering stage' of spiritual training.

Before the essence can be perceived by the awareness soul there is one further stage at the level of the intellectual and mind soul. To get an idea of a tree, Steiner said, one may 'take photographs of it from all angles;' 'the greater the number of photographs, the closer will the image I have of it get to the reality of the tree.' We need different aspects of the tree, but these must not be superimposed on each other on the photographic plate. It is important 'to keep the different aspects apart',[12] otherwise the images in our minds become blurred and the concepts we form lack clarity. Looking back to the way organs are formed from concepts in a process that occurs in unconscious soul life and in the ether body, we perceive enhancement: individual concepts become individual aspects that are the product of assimilation in thought. Will these, too, give rise to a new organ?

The intellectual element in the intellectual and mind soul keeps individual aspects apart and also keeps them together, compares one with another, and thus unites them. The sentient soul is also required to take up every aspect in a living way, so that the mind soul may receive it in a more inward way. The ego lives in a breathing process that goes to and fro between world and the inner soul in the intellectual and mind soul; it is able to let individual aspects gained from the world follow one another in sequence in the soul, or make one or the other the focus of awareness. It needs the awareness soul, however, to hold an aspect fast and transform it.

Awareness Soul and Spiritual Training

So far we have presented the preparatory, introductory stage of spiritual training. Again it is possible to stop at this, which will once more lead to the unhappy experience that whatever we have gained cannot be held on to but only developed further. Unless we make continuous efforts–also with anything we experienced on reading a book–to feel our way to the reality of the spirit, we do not progress beyond reflecting on contents mediated through spiritual science and relating to them with mind and soul, and we feel how their life seeps away. Again it is necessary to look back and ask oneself: When did I succeed in developing real presence of spirit in my thinking; when were my actions governed by spiritual points of view? When it comes to spiritual training, we are still only at the beginning of the awareness soul epoch.

Spiritual training as such can only be achieved through the

awareness soul. It takes its beginning in the alertness of this soul element, enhancing and extending it. So far we have noted the relationship between individual soul elements and spiritual development. At this point, however, the development of the awareness soul unites with the training process, and the development of the awareness soul towards the spirit now *is* the training process. The intellectual and mind soul creates the space for meditative life in the soul, a space filled with the active mindfulness of the awareness soul. In the process of meditation this soul element concentrates all powers of soul on the object it has made the centre and lets the image of the object become transparent, so that the aspect held in focus in the soul becomes a window through which a ray of eternity may be received in full awareness by the soul.

The training establishes a new spiritual relationship to the world that does not stop at seeing various aspects of a particular object. At the centre of all the different views we have obtained of a tree stands the tree itself in its spiritual reality, something that so far could only be approximated. The concept 'tree' pointed to it, and living thoughts were feeling their way towards it. Now our perception of the reality, which so far has been more or less shadowy, is enhanced. The awareness soul has an experience of manifestation as some of the essence of the object reveals itself at the centre of the awareness soul, at the centre of the different aspects that have become organ, and this may be like a sudden flash of lightning, or like a light that gradually illumines. The sentient response is: That is the reality, and in fact the only reality there is. Projective thought fed by this light can then let the tree come into being again, even if initially this can only be done to a minor degree.

The above has nothing to do with the soulful mystical experiences that go hand in hand with a reduced state of conscious awareness. It is living experience gained through thinking activity in enhanced awareness. In his *Philosophy of Freedom*, Steiner called the form in which the thought content arises in the soul 'Intuition'. 'In the living experience that is developed in thinking activity, this Intuition is able to enter to a greater or lesser extent into the realities behind the realities.'[13] This indicates that the awareness soul establishes stages of intuitive thought activity for spiritual training. The will of the awareness soul becomes creative in our thinking when we are involved in intuitive activity of this kind, and this leads to the experience of essences becoming manifest. The 'characteristic element of Intuition is also one of manifestation, of the essence being open to insight and transparent' (Witzenmann).[14]

Something of the light of the awareness soul may show itself in the soul even when we first begin to work with contents presented by spiritual science.

When there has been this first 'contact of the soul with the supernatural world,' images rise 'from the waters of soul life' that are 'wrought entirely by the soul itself'. They are 'like a curtain that the soul puts in front of the supernatural world' when it 'feels itself touched by that world'.[15] Our own creative etheric powers provide the 'substance' for those images, i.e. for their composition and their organs, as already mentioned, and the images may be felt to provide protection against being overpowered by the spiritual experience and to be a first attempt to assimilate it.

Schiller drew attention to another aspect. 'The purely spiritual experience is certainly there for the disciple, but his capacities are not yet sufficiently developed to experience it in awareness in its purely spiritual form. To begin with, his awareness is filled only with the image that is made up of mental images from the ordinary life of the soul, an image wrought in colour, form and movement.'[16] We know that behind it stands the spiritual reality, a glimmer of which had already shown itself in the soul through Intuitive thought activity. Now it has grown richer and comes to expression in images.

Thus we see that there are also stages in the path to 'clairvoyant Imagination', stages that lead to the point where the threshold of the spiritual world is crossed and represent 'the first steps towards Imaginative perception'. The disciple becomes aware of this because the world of abstract thought 'is imbued with inner life.'[17] We are on the way to this even when we try to let the concepts of spiritual science become images, thus making use of the power of imagination in a way that relates to reality.[18] Steiner tells us not to despair. 'Even if the images are not the right ones to begin with, this does not matter. They will be corrected by those that guide us.' This of course does not free us from the obligation to combine the courage to form images with making every effort to make those images as accurate and as true as possible.

Yet wisdom transformed into an image can heal. The Imagination 'penetrates right to the ether body' and this mediates the restoration of health.[19] People hunger to perceive images today, and the quick and easy gratification of this hunger is sought through drugs. Surely this is the hunger for Imagination, and expresses a longing for the world of the spirit that initially wants to come to revelation in this form? The images that arise on the road to Imagination can satisfy that hunger in a way that is appropriate to the present age and can bring healing. It does however call for spiritual activity and alertness.

Imaginative perception initially beholds the individual's 'own inner world'; later 'images of the cosmos'[17] arise where one no longer has the feeling that something is merely coming up from below but rather that something is coming in. These are Poppelbaum's 'etheric token images', and in them, the striving for truth that fills the intellectual and mind soul finds its first fulfilment through images in the sphere of the awareness soul.

This living Imaginative experience in the awareness soul brings a 'feeling of tremendous subjective bliss' that initially arises through enhancement of inner activity, the 'active living experience of one's own individual personality'.[20] The feeling of pleasure that comes with every experience connected with the ether body no doubt also plays a role. At a higher level, bliss comes with the liberating experience one gains when 'expansion into the universal ether', into the etheric world, happens from the centre, the 'ego point' of the awareness soul. The life in two-dimensional images that belongs to the etheric region, a life in which we become active ourselves, where we move as we feel our way, can leave the third dimension of space that we know on earth behind.[20] The feeling of bliss that arises does not come from but streams into the awareness soul. This feeling can present a danger to the disciple, but help is given in so far as the awareness soul always demands that we are fully aware that the images received through Imagination are not yet the reality of the spirit but merely images of it.[17]

Transformation of the Awareness Soul

Living in expansion, the awareness soul grows beyond itself. Striving towards the higher ego and the world of the spirit, it also seeks a new relationship with the earthly world, a relationship in spirit. Due to the essential nature of the awareness soul this causes problems in social life.

It has been shown that in the second half of life an attitude where one is largely taking from life gradually gives way to one where one is purely giving, and the flowering of the awareness soul wants to progress to the bearing of fruit. That, however, is not something inherent in the awareness soul as such. This soul is essentially antisocial rather than social. Steiner even spoke of 'antisocial instincts' that would of necessity arise in the epoch of the awareness soul. Wide-awake alertness is not in itself something that creates a bond between people; our nature is such that we really are social creatures only when we are asleep.[21] To achieve full awareness and develop their own essential nature, human beings need first of all to establish their identity as distinct from others. This also applies to human communities, which should provide not only communal life but also a distinct identity. If this does not happen, or if there is not enough of this, the previously discussed problems in the development of the inward-turned intellectual and mind soul may arise. Even greater difficulties arise when it comes to the development of an awareness soul that becomes creative out of inner resources. Young people in particular exhaust themselves if they cannot develop an adequate inner life.

How can the awareness soul grow beyond itself and become social and part of a new community? By first of all creating a 'social structure' based on spiritual perception; with this structure it is then possible to 'tame the instincts of antisocial images'.[21] This is the call for a human society based on spiritual principles, and in the final instance the 'threefold social order'[22] that is also demanded by the developmental problems that arise for the awareness souls of individuals.

How can individuals endow such a structure with life? Social life is initially governed by antisocial instincts. One wants to keep a distinct identity and this may lead to antipathy and the urge to criticize, with the result that development regresses to the level of the intellectual and the sentient soul. At the same time, a positive attitude is practised at the level of the awareness soul, and this is one of the basic exercises in spiritual training. I want to see the positive side of this individual! Such a resolution may remain abstract and not go beyond the head. If it is nevertheless carried out, the whole thing seems forced. The positive attitude, and even the tolerance that precedes it, does not come from the heart and therefore seems superficial.

Here Steiner spoke of the help that comes when the awareness soul grows beyond itself and thus takes us forward. He suggested that we should deliberately increase our interest in the other person and create images of the other person, so that we get away from generalizations in which a person is considered either good or bad. This will help us to overcome not only antisocial antipathies but also sympathies that appear antisocial because they exclude people with whom we are not in sympathy from social contact. The image lets us take in something of the other person without having to break down the necessary boundaries; it leaves us free but offers the opportunity to enter into a different relationship with the other person.

This has nothing to do with making abstract lists of the good and bad points of the other person in form of images. Steiner speaks of images that come as we 'ascend to Imaginative life'.[23] These are token images in which we aim to reach the positive core of the other person. It will then no longer be necessary to force ourselves to take a positive view, because we experience, or at least glimpse, the positive element in the other person. We may start with the outer appearance of the other individual, but his words and actions and the effect he has in the social sphere must also be included. All kinds of aspects are considered as we feel our way to the essential image. And even as we feel our way, a feeling of fairness towards the other person begins to develop, and this can then radiate into earthly existence.

Growing beyond itself, the awareness soul undergoes a transformation. 'Emphasis on the ego steps into the background; instead, the awareness soul, which formerly served mainly to cultivate the ego, gradually fills up with what we call Imagination'.[24] The relationship between the awareness soul and the brain becomes freer, but such a fundamental relationship nevertheless still exists in the process of Imagination—via the liberated ether body of the brain. 'In the brain, nature presents as a real Imagination perceptible to the senses what in fact is achieved in a higher sphere through Imaginative perception.'[12] The brain is the physical, the Imagination the nonphysical image of a spiritual reality; what they have in common and what unites them is the image-forming activity.

In the further course of soul development, with development of the spirit rising out of it, 'the awareness soul is transformed into the Imagination soul'.[24]

Transformation of the Intellectual and Mind Soul

As the awareness soul may become the Imagination soul, so the intellectual and mind soul may become the Inspiration soul.[24]

In Inspiration, the spiritual world no longer presents itself in images; its essence and the entities within it now 'speak' to human beings.

In the process of Imagination, thinking is enhanced to become spiritual vision; in Inspiration, feeling is enhanced to spiritual hearing. To begin with, the disciple clears the inner image he has entered into so deeply from his conscious mind, letting it grow quite empty. He must inwardly 'feel' everything he did to produce the image.[25] Human beings have the latent faculties for Inspiration just as much as for Imagination. Clairvoyant Inspiration 'merely means to bring into the light of awareness the Inspirations that are already there in the feeling life of every individual, though at an unconscious level.'[26]

Compared to the bliss felt in the process of Imagination, it causes 'profound pain in the soul' to let go of images that have become dear to us and to let the conscious mind grow empty. Disciples come 'to know the truth that the whole of existence must in the final instance be born out of pain.' (This is also evident from the many birth pangs in the biography.) Emptiness, the first precondition for Inspiration, is then joined by a silence 'more silent than silent' that has to be taken through the zero point. The world of the spirit speaks when 'hearing has become negative'.[17]

Steiner compared the process to inhalation, 'inspiration' being a synonym of this. This opens up new insight into the human breathing process, and Steiner actually called 'everything that is connected with breathing, Inspiration brought to realization in the physical world.'[12] Schiller sums it up as follows: 'Air enters into the lung once this has been emptied in the process of exhalation. Inspiration means that the essence enters into a soul that has been deliberately made empty. ... We exhale and return to earthly life.'[27]

With Inspiration, we enter into the atmosphere of an intellectual and mind soul that is becoming spiritual. Once again we experience the breathing that goes to and fro between the inner soul and the world, but this time it is the world of the spirit to which the soul relates in a breathing process. We experience a life of feeling that is growing spiritual, falling into silence and emptying itself, but this time not for another individual with whom we have a connection in mind and soul, but for the entities in the world of the spirit. The intellect of this soul must forego 'independent thought' so that after also letting go of our 'ordinary life in mind and soul' Inspiration may enter.[24]

This does not mean that we should forego independent thought and grow impoverished in mind and soul in our earthly lives. It is rather that intellect and mind become instruments for a higher life, one that human beings embody in individual form. When we return to earthly life from our spiritual researches we will still need practical independent thinking and a warm heart, though these will now be irradiated with the fruits of our spiritual labours. This is something that could be experienced in the personality of Rudolf Steiner.

The social consequences will go beyond those that came when the awareness soul entered into Imagination. Social Inspirations may develop and bring something new into the social sphere. And the transformation of the intellectual and mind soul also proceeds in stages that gradually take one in the direction of clairvoyant Inspiration. Steiner said, 'The works of inspiration that have become part of the cultural heritage have been inspired in a transformed intellectual soul.'[24] The preliminary stages can also be experienced in exchanges from spirit to spirit and soul to soul between individuals, where one person may 'inspire' the other.

Transformation of the Sentient Soul

Intuition directly unites the ego with the entity that has 'spoken' to it through Inspiration. Now the 'element of will becomes an organ of perception'.[28] An act of will at the level of ordinary life may be taken as the model: total immersion of soul and spirit in the metabolic system of the organism. In the world of the spirit this becomes the highest form of perception. Steiner said with reference to these will dynamics, 'That, in essence, is the nature of Intuitive perception.'[12] Before there can be such perception, the disciple must again let go of things that he had before. Having let go of the images, he is now asked to let go of 'living in his own activity of soul into which he had entered so deeply in order to acquire the faculty for Inspiration.'[29]

In Intuition, the will becomes the organ for perceiving the reality of another entity. But it is the power of a love grown spiritual that gives life to the union with that entity. The power of love becomes 'power of perception'. We have seen how in love the individual reality of the individual comprehends itself and at the same time takes the reality of the other into itself. Now love becomes the pure emanation of the higher ego; this unites completely with the other entity, yet as in every other kind of love, does not lose its identity. Intuition is achieved 'when we enter into living experience of something other, remaining totally individual yet also in utter selflessness.'[28]

Just this one aspect: 'Soul and spirit becoming totally immersed when there is an act of will–highest form of perception', immediately points to the tremendous range that there is to this process. This is also borne out by Steiner's comment that the first and lowest soul principle, the sentient soul, can serve the higher form of perception, which is Intuition, and be transformed into the Intuition soul in the process. On the other hand the sentient soul is not only the lowest soul element but also 'the richest soul'. The elemental powers of its 'inner impulses, inner passions and affects that act as will-determined drives may be transformed into Intuitions.'[24] In the process of Intuition the immediacy of sentient experience of the world becomes an immediacy of uniting one's essence. The sentient soul opened up to the world. Its transformation begins when individuals 'make world interests their own and thus grow increasingly beyond personal sentience.'[24] The foundations for this may be laid in the twenties.

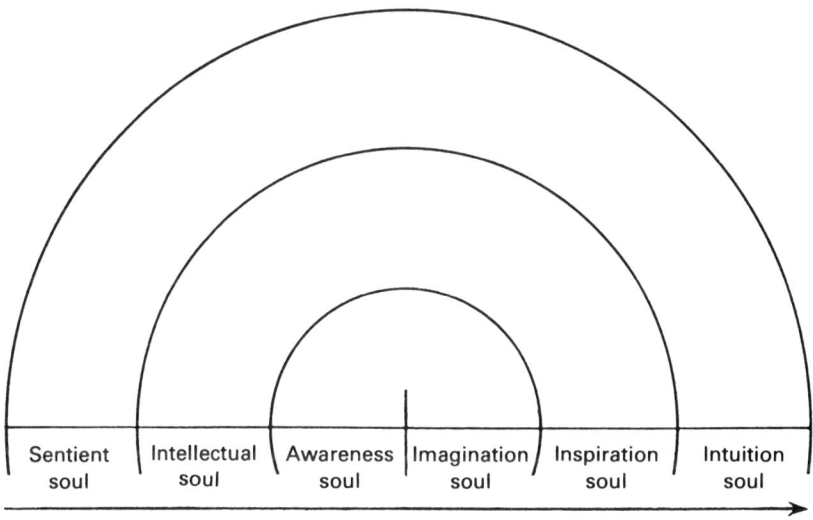

| Sentient soul | Intellectual soul | Awareness soul | Imagination soul | Inspiration soul | Intuition soul |

These considerations show that there are also stages of Intuition, and Steiner put great emphasis on this. 'In their moral sense, in the contents of their moral conscience, human beings have an earthly reflection of what happens in the process of Intuition;'[17] or: 'Even the simplest thought has an element of Intuition.'[30] This last aspect relates to the concept of Intuition in *The Philosophy of Freedom*. A further, wide-ranging connection thus exists between Intuitive perceptions made as the thinking of the awareness soul expands to encompass the essence, perceptions that shine forth even before Imagination is achieved, and Intuition at the third level of perception.

The Biography of Spiritual Training

The above diagram shows the biography of spiritual training. This training can bring progress in soul development and at the same time also a transformation of the soul principles. Our own ego is to an increased degree determining the biography of spiritual training, and this means that this biography takes a more individual form that the earthly biography. How much is achieved how soon depends largely on the disciple. According to Steiner, it is not possible to make general statements as to the duration of different stages. And it is even more the case than it was with soul development, that there is no rigid sequence, with one stage following the other. It has been shown that the untransformed soul principles are also involved in the process of spiritual training, though the emphasis is always on one particular principle. Steiner said that some individuals were able to practise 'Inspirative and Intuitive perception almost simultaneously with Imaginative perception'. 'This should not, however, be taken to mean that anyone might be spared the process of having to go through Imagination.'[31] 'For Western civilization, the path to the supernatural worlds' is the path of Imagination.[32]

If no account is taken of this and people aim for Inspiration or Intuition before they have developed Imagination, Steiner said that 'it may happen that someone has only developed as far as the sentient soul, i.e. that he gives free rein to his personal desires, drives etc. Let us assume such a person had become wound up by going through occult development. The consequence would be that his sentient soul was transformed into an Intuition soul and he would have certain Intuitions; these, however, would be nothing more than his own personal drives, desires and instincts presenting themselves in a different form. An individual who has undergone moral development to the level of the intellectual soul, i.e. has acquired clear-cut and more general concepts, and whose mind encompasses general world interests ... will transform at least his mind soul into Inspiration soul and achieve certain Inspirations, though his clairvoyant powers will not be entirely purified. It is not until human beings have truly penetrated as far as the awareness soul with their egos that they achieve what initially is the transformation of their awareness souls into Imagination souls.' The other transformations can then follow and be 'perfectly safe',[24] because the ego of the awareness soul will watch over every transformation. This points to the mid-life period; from then on the soul principles can be rightly transformed. Steiner altogether called this 'the best time at which to develop one's spiritual faculties',[33] a process that can then continue as life goes on. However, the awareness soul casts its light on the whole of soul development; exercises for the attainment of Imagination, Inspiration and Intuition may therefore begin in the twenties. Imaginative images can arise when the awareness soul is still in its germinative stage.

For anthroposophical physicians who want to take their first steps in spiritual training, the sequence of those steps is important. Steiner charged physicians to explore the inner world of the organism and of diseased organs by means of Imagination, and the kingdoms of nature and the medicines obtained from them through Inspiration. For therapy Intuition unites Imagination and Inspiration.[34] The prior stage of Imagination is not specifically mentioned when it comes to gaining insight into the kingdoms of nature and of medicines, but it is evident from the above that it, too, is required.

Physicians will first of all try and form images of diseased organs and of medicinal plants and substances, images that go towards Imagination. To discover the living activity of a medicine, or rediscover an existing medicine in this way, they must let go of the images and listen to what the plant or the substance has to tell them. Subtle feelings may be experienced in relation to potential medicines, a mood that Steiner compared to a musical experience when speaking of Inspiration.[35] The image of the disease, recreated anew, may be experienced as a silent question, the medicine as the answer that 'sings' inside. This brings to mind the words of Novalis, that 'every illness is a musical problem, and healing the musical resolution.'[36]

When the will to heal comes to life out of this mood it adds another element: For just a moment, the physician feels as if united with the patient and his recovery. He makes the Intuitive will-based decision: This is the medicine for this particular patient, in this potency and this particular formulation. Such starlit hours may be rare, and they may only mark the preliminary stage to the research and therapy of the future, but they are sufficiently powerful to spread their radiance over the physician's everyday work, letting something of the experience that filled them remain.

However individual our spiritual training may be, and even though we only achieve it in part, certain laws pertain also to its biography. It has already been said that wanting to linger at a particular stage is detrimental to spiritual development. On the other hand there may be a desire to skip a stage or two in spiritual training. The dangers that may arise from this have been discussed (withering in the light of conscious awareness). In the light of the above, however, other consequences may also arise. If the stage of the sentient or the intellectual and mind soul is neglected in spiritual training, the disciple lacks a sound basis not only for the development of the awareness soul, but also for the transformation of the other two soul principles into Inspiration and Intuition souls. To help this development as it begins to take place, the sentient soul and the intellectual and mind soul need to be revitalized over and over again for the rest of the disciple's life.

The laws pertaining to the biography of spiritual development also show that there is a direct relationship to the earthly biography. 'In a sense' disciples of the spirit are also 'dependent on their age, or life stage.' For perception of the supernatural, his 'life stages become organs of comprehension' through which Inspirations become possible. His Inspirations will be different, and less complete, at a younger age. He will have to wait until he has reached a certain age before he can explore a particular thing.[37] Spiritual development may receive new impetus particularly in the early forties. Apart from the transformation of the soul principles, and this may already be initiated during soul development, the opportunity now arises to begin with the spiritual transformation of the principles involved in physical development, initially the astral body.[38]

A theme thus shows itself in conclusion that is present throughout both the earthly biography and the biography of spiritual training. We must learn to wait and exercise patience. Impatience inhibits or prevents progress on any path of human evolution. We have to wait to be born into this world, or we ought to wait; in the same way we must be able to wait for the birth processes that occur in soul and spiritual development, having done everything necessary in preparation. It may then be experienced as a moment of grace when something new wants to come into life from the world of the divine, something to give new impetus to development, and healing in case of illness.

Bibliography

Chapter 1

1 Hiebel, F. *Biographik und Essayistik*, S. 159. Bern/Muenchen 1970.
2 See Poppelbaum, H. *Mensch und Tier*, S.99. Basel 1928.
3 Part of Goethe's poem 'Vermaechtnis'.
4 Steiner, R. *Theosophy* Tr. A.P. Shepherd. London: Rudolf Steiner Press 1973.
5 Klages, L. *Vom Wesen des Rhythmus*, Kampen/Sylt 1934.
6 Steiner, R. *Metamorphoses of the Soul. Paths of Experience* Vol..2. Tr. C. Davy and C. v.Arnim. London: Rudolf Steiner Press 1983.
7 Hoerner, W. *Zeit und Rhythmus*, S.92. Stuttgart 1978.
8 ibid., S.112.
9 ibid., S. 177.
10 Treichler, R. Lebensstufen und Seelenentwicklung. Im *Anthropo. Mediz. Jahrbuch*, Bd.II, Dornach 1951.
11 Buehler, W. *Das bewegliche Osterfest*, S.22. Tuebingen 1965.
12 Thomae, H., der W. Hellpach referiert, in *Psychologische Probleme des Erwachsenenalters*, S. 688 ff. Universitas, Juni 1957.
13 Steiner R. *The Education of the Child in the Light of Anthroposophy* Tr. M. and G. Adams, London: Rudolf Steiner Press 1975.
14 Guardini, R. *Die Lebensalter*, S. 11. Wuerzburg 1959.
15 See Lievegoed, B. *Lebenskrisen-Lebenschancen*, S.33/34, Muenchen 1979.
16 See Hessenbruch, H. *Von der Bedeutung des Siebenjahresrhythmus beim heranwachsenden Menschen*. Dissertation 1938, S.3.
17 Moers, M. *Die Entwicklungsphasen des menschlichen Lebens.*. Ratingen 1953
18 Kuenkel, H. *Die Lebensalter*, S. 28 ff. Braunschweig 1948.
19 Sheehy, G. *In der Mitte des Lebens*, S. 33. Frankfurt/M. 1978.
20 ibid., S.19.
21 ibid., S. 381.
22 Buehler, C. *Der menschliche Lebenslauf*. Leipzig 1933, Neuaufl. Goettingen 1959.
23 Oerter, R. *Moderne Entwicklungspsychologie*, S. 15. Donauwoerth 1976.
24 Schraml, W. *Einfuehrung in die moderne Entwicklungspsychologie*, S. 14. Stuttgart 1972.
25 Hahn, H. *Der Lebenslauf als Kunstwerk*. Stuttgart 1966

Chapter 2

1 Steiner R. *The Education of the Child in the Light of Anthroposophy* Tr. M. and G. Adams,London: Rudolf Steiner Press 1975.

2 Seitelberger, F. *Handbuch der Kinderheilkunde*, VIII, 1, S.48. Berlin 1969.

3 Husemann, F./Wolff, O. *Das Bild des Menschen als Grundlage der Heilkunst*, Bd. II, 1, S. 22. Stuttgart 1974.

4 Steiner, R. *Theosophy* Tr. A.P. Shepherd. London: Rudolf Steiner Press 1973.

5 Buehler, W. *Der Leib als Instrument der Seele*, Stuttgart, Aufl. 1976.

6 Steiner, R. *The Study of Man. General Education Course.* Tr. D. Harwood and H.Fox. Lectures of 30.8. and 3.10.1919. London, Rudolf Steiner Press 1975.

7 Treichler, R. Das Volksschulalter in aerztlicher Sicht. *Erziehungskunst*, S.312, Juli 1957.

8 Driesch, H. *Philosophie des Organischen*, S. 139. Stuttgart 1921.

9 Steiner, R. *Theosophy* Tr. A.P. Shepherd. London: Rudolf Steiner Press 1973.

10 Steiner, R., Wegman, I. *Fundamentals of Therapy* Tr. E.Frommer and J. Josephson. London: Rudolf Steiner Press 1983.

11 Steiner, R. *Spiritual Science and Medicine* Lecture of 27.3.1920. London: Rudolf Steiner Press 1948.

12 Steiner, R. *At the Gates of Spiritual Science* Tr. E. Goddard and C. Davy. London: Rudolf Steiner Press 1970.

13 Vogel, L. *Der dreigliedrige Mensch*, S. 215. Dornach 1967.

14 Steiner, R. *The Wisdom of Man, of the Soul and of the Spirit* tr. S. and L. Lockwood. Lecture of 23.10.1909. New York: Anthroposophic Press 1971.

15 ibid. Lecture of 25.10.1909.

16 Steiner, R. *The Study of Man. General Education Course.* Tr. D. Harwood and H.Fox. Lecture of 28.8.1919. London, Rudolf Steiner Press 1975.

17 *Goethes Naturwissenschaftliche Schriften* Herausgeg. durch R. Steiner. Bd.II, S.32. Bedeutende Fördernis durch ein einziges geistreiches Wort.

18 See *Vom Lebenslauf des Menschen* Themen aus dem Gesamtwerk Rudolf Steiners, S. 239. Stuttgart 1980.

19 Koenig, K. Die menschliche Seele, S. 125 ff.; in: *Aspekte der Heilpaedagogik.* Stuttgart 1969.

19a Steiner, R. *Das Verhältnis der verschiedenen naturwissenschaftlichen Gebiete zur Astonomie.* Lecture of 2.1.1921. Ms R81, Rudolf Steiner House Library.

20 Matthiolius, H. Temperamente; in: Husemann/Wolff: *Das Bild des Menschen* ..., S.80, Bd III, 1; vgl. Ref. 2:3.

21 Holtzapfel, W. *Krankheitsepochen der Kindheit*, S.37. Stuttgart 1978.

22 Steiner R. *Curative Education* Tr. M. Adams. Lecture of 28 6.1924. London: Rudolf Steiner Press 1972

23 Steiner, R. *Soul Economy and Waldorf Education* Lecture of 31.12.1921. London Rudolf Steiner Press 1986.

24 Mueller-Wiedemann, H. *Mitte der Kindheit*, S. 104 ff. Stuttgart, Aufl. 1980.

25 Steiner, R. *Von Seelenrätseln* Ausg. 1917, Kap. IVb, S. 235. GA 21.

26 Ausfuehrlicher bei R. Treichler; Grundzuege einer geisteswissenschaftlich orientierten Psychiatrie, in: Husemann/Wolff: *Das Bild des Menschen...*, Bd II,2.

27 Steiner, R. 'The spiritual-scientific aspect of therapy'. Lecture of 14.4.1921. Manuscript translation. Association of Anthroposophical Doctors.

28 Holtzapfel, W. *Krankheitsepochen der Kindheit*, S.43 ff.; vgl. Ref. 2:21.

29 Jung, C.G. *Seelenprobleme der Gegenwart*, S.180/181. Olten/Freiburg 1973. Zum Jugendalter vgl. ferner: E. Spranger, *Psychologie des Jugendalters*. Heidelberg, Aufl. 1966.

30 Steiner, R. *Soul Economy and Waldorf Education* Lecture of 4.1.1922. London Rudolf Steiner Press 1986.

31 Thomas, K. *Abriss der Entwicklungspsychologie*, S. 178 ff. Freiburg 1978.

32 Steiner, R. *Spiritual Science and Medicine* Lecture of 8.4.1920. London: Rudolf Steiner Press 1948.

33 Leber, S. *Geschlechtlichkeit und Erziehungsauftrag*. Stuttgart 1981.

34 Steiner R. *Curative Education* Tr. M. Adams. Lecture of 25 6.1924. London: Rudolf Steiner Press 1972

35 Husemann/Wolff, *Das Bild des Menschen...*, Bd II,1, S. 45 ff; vgl. Ref. 2:3.

36 Steiner, R. *Soul Economy and Waldorf Education* Lecture of 4.1.1922. London Rudolf Steiner Press 1986.

37 Wolff, O. Das Nierensystem; in: Husemann/Wolff, II,2, S. 537; vgl. Ref. 2:3.

38 Steiner, R. *Supplementary Course-The Upper School*. East Grinstead: Michael Hall 1965.

39 Steiner, R. *Friedrich Nietzsche*. Tr. M. Ingram deRis. New Jersey: Rudolf Steiner Publications 1960. Koenig, K. *Die menschliche Seele*, S. 104ff. This also has the quote from Heidegger.

40 Steiner, R. *Rhythmen im Kosmos und im Menschenwesen*. Lecture of 6.6.1923. GA 350.

41 Steiner, R. *Man's Being, His Destiny and World Evolution.*Tr. E. McArthur. Lecture of 20.5.1923. New York: Anthroposophic Press 1966.

42 Steiner, R. 'Seven lectures to workmen'. Tr. M. Cotterell. Lecture of 2.6.1923. Manuscript translation R 64. London: Rudolf Steiner House Library. Concerning the importance of imagination, see also Fucke, E. *Die Bedeutung der Phantasie*. Stuttgart 1972.

43 Fina, K. Zit. n. Mueller-Wiedemann: *Mitte der Kindheit*, S. 119; vgl. Ref. 2:24.

44 Mueller-Wiedemann, *Mitte der Kindheit*, S. 152; vgl. Ref. 2:24.

45 Freud, S. *Vorlesungen zur Einführung in die Psychoanalyse* I S. 48 under 365f. Ausg. 1969 Frankfurt a.M.

46 Steiner, R. *Geisteswissenschaftliche Behandlung sozialer und pädagogischer Fragen*. 9. Vortr., GA 192.

47 Steiner, R. *The Philosophy of Freedom*. Tr. M. Wilson. Ch. 4 and 5. London: Rudolf Steiner Press 1970.

48 Zeylmans van Emmichoven, F.W. *Die menschliche Seele*, S. 107. Basel 1953.

49 Lindenberg, C. Vorstellen, Begehren und Urteilen; in *Die Drei* 9/1976; S.434.

50 Steiner, R. *The Wisdom of Man, of the Soul and of the Spirit* tr. S. and L.

Lockwood. Lecture of 26.10.1909. New York: Anthroposophic Press 1971.
51 Lindenberg, C. Op.cit. S.439.
52 Steiner R. *Erziehung und Unterricht aus Menschenerkenntnis*. 5. Vortr. 21.6.1922. GA 302a.
53 Zeylmans van Emmichoven, Op.cit. S.32.
54 Steiner, R. *Theosophy* Tr. A.P. Shepherd. London: Rudolf Steiner Press 1973.
55 Steiner, R. *Occult Science. An Outline*. Tr. G. and M. Adams. London: Rudolf Steiner Press 1969.
56 Steiner, R. *Metamorphoses of the Soul. Paths of Experience* Vol..2. Tr. C. Davy and C. v.Arnim. London: Rudolf Steiner Press 1983.
57 Buehler, C. *Der menschliche Lebenslauf*, S. 206 ff.; vgl. Ref. 1:22.
58 Glas, N. *Das Antlitz offenbart den Menschen*, S.16. Stuttgart 1963.
59 Zit. n. Kretschmer, E. *Medizinische Psychologie*, S.8. Stuttgart 1975.
60 Buehler, C. *Psychologie im Leben unserer Zeit*, S. 215 u. 245. Stuttgart/Hamburg 1962.
61 Jung, C.G. *Die Beziehungen zwischen dem Ich und dem Unbewussten*, S. 98 u. 206. Darmstadt 1938.
62 Steiner, R. 'The ego as experience of consciousness'. Lecture of 20.8.1921. Manuscript translation R 66, London, Rudolf Steiner House Library. *The Gospel of St John*. Tr. M.B. Monges. Lecture of 31.5.1908. New York: Anthroposophic Press. *The Secrets of the Threshold*. Lecture of 29.8.1913. London: Anthroposophical Publishing Co. 1928.
63 Steiner, R. *Knowledge of the Higher Worlds*. Tr. G. Metaxa. London: Rudolf Steiner Press 1976.
64 Husemann, F. Der Mensch und sein Genius, *Vortragswerk 1*, 1962.
65 Ausfuehrlicher in: Treichler, R. Schlafen und Wachen als Atmung des Ich; in: *Beitraege zu einer Erweiterung der Heilkunst*, 1/1970.
66 V.d.Heide, P. Das Wesen des Ichs; in: *Therapie seelischer Erkrankungen aus anthroposophischer Sicht*, S. 39 ff. Stuttgart 1979.
67 See also Fromm, E. *Die Kunst des Liebens*, S. 39 ff. Frankfurt/M. 1979.
68 Erikson, E. *Identitaet und Lebenszyklen*, S. 124 u. 107. Frankfurt/M. 1979.
69 Lievegoed, B. *Lebenskrisen ...*, S. 50 ff; vgl. Ref. 1:13.
70 Glas, N. *Jugendzeit und mittleres Lebensalter*, S. 6 ff. Stuttgart 1960.- Lauenstein, D. *Der Lebenslauf und seine Gesetze*, S. 56 u. 66 ff. Stuttgart 1974.
71 Hoerner, W. *Zeit und Rhythmus*, S. 209 ff.; vgl. Ref. 1:7.
72 Steiner, R. *Karmic Relationships* vol. 1. Tr. G. Adams. Lecture of 16.2.1924. London: Rudolf Steiner Press 1972.
73 Lievegoed, B. Op.cit. S. 68 ff.
74 Buehler, C. *Psychologie ...*, S. 264; vgl. Ref. 2:60.
75 Hahn, H. *Der Lebenslauf als Kunstwerk*, S. 90; vgl. Ref. 1:25.
76 Kipp, F. *Die Evolution des Menschen im Hinblick auf seine lange Jugendzeit*. Stuttgart 1980.
77 Steiner, R. *The Gospel of St John in Relation to the Other Three Gospels*. Tr. S. and L. Lockwood. Lecture of 25.6.1909. New York: Anthroposophic Press 1982.
78 Steiner, R. *Metamorphoses of the Soul. Paths of Experience* Vol..2. Tr. C.

Davy and C. v.Arnim. London: Rudolf Steiner Press 1983.
79 Steiner, R. *Occult Science. An Outline.* Tr. G. and M. Adams. London: Rudolf Steiner Press 1969.
80 Steiner, R. *The Wisdom of Man, of the Soul and of the Spirit* tr. S. and L. Lockwood. Chapter 4. New York: Anthroposophic Press 1971.
81 Kuenkel, H. *Die Lebensalter,* S. 56; vgl. Ref. 1:18.
82 Steiner, R. *Health Care as a Social Issue.* Lecture of 7.4.1920. Mercury 1984.
83 Sheehy, G. *In der Mitte des Lebens,* S. 143 ff.; vgl. Ref. 1:19.
84 Zu den Jahrsiebten in Goethes Leben vgl. Hiebel, F. *Goethe.* Berlin 1961.
85 Kuenkel, H. Op.cit. S. 67.
86 Steiner, R. *Theosophy* Tr. A.P. Shepherd. London: Rudolf Steiner Press 1973.
87 Steiner R. *The Education of the Child in the Light of Anthroposophy* Tr. M. and G. Adams, London: Rudolf Steiner Press 1975.
88 Steiner, R. *Menschliche und menschheitliche Entwicklungswahrheiten.* 1. Vortr. GA 176.
89 Steiner R. *Vom Lebenslauf des Menschen,* S. 159; see Ref. 2:18.
90 Buehler, C. *Der menschliche Lebenslauf,* S. 209.
91 Sheehy, G. op.cit., S. 251; vgl. Ref. 1:19.
92 Vischer, A.L. *Seelische Wandlungen beim alternden Menschen,* S. 215. Basel 1944.
93 Buehler, C. *Psychologie* , S. 215; vgl. Ref. 2:60.
94 Quoted from Buehler, C. *Der menschliche Lebenslauf,* S. 280. Vgl. Ref. 1:22.
95 Ibid. S. 301.
96 Steiner, R. *Goethe's World View.* Obtainable via London: Rudolf Steiner Press Mail Order.
97 Steiner, R. Ibid.
98 Schiller, F. *Ueber die aesthetische Erziehung des Menschen,* 12., 14. u. 18. Brief.
99 Zu diesen Beispielen vgl. Lauer, H.E. *Vom richtigen Altwerden,* S. 93 ff. Freiburg 1972.
100 Steiner, R. *Earthly and Cosmic Man.* Tr. D.S. Osmond. Lecture of 23.4.1912. London: Rudolf Steiner Publishing Co. 1948.
101 Buerger, N. *Pathologische Physiologie,* S. 619. Leipzig 1956.
102 Wolff, O. in Husemann/Wolf, I, S. 131; vgl. Ref. 2:3.
103 Holtzapfel, W. *Krankheitsepochen der Kindheit,* S. 13 ff.; vgl. Ref. 2:28.
104 Steiner, R. *Philosophie und Anthroposophie.* S. 415. GA 35.
105 Hampe, J.C. *Sterben ist doch ganz anders,* S. 12. Stuttgart 1975.
106 Steiner, R. *Metamorphoses of the Soul. Paths of Experience* Vol..2. Tr. C. Davy and C. v.Arnim. London: Rudolf Steiner Press 1983.
107 Steiner, R. *Aus dem mitteleuropäischen Geistesleben.* Vortr. 13.4.1916. GA 65.

Chapter 3

1 V.Schroetter, H. Die Persoenlichkeit des Infantilen; in: *Infantilismus. Psychologische Praxis,* S. 43, Heft 16. Basel 1955.
2 Heymann, K. ibido, S. 9.
3 Thomae, H. ibido, S. 40.

4 Kretschmer, E. *Medizinische Psychologie*, S. 148 u. 163. Stuttgart 1975.

5 Winkler, W.T. in *Infantilismus*, S. 83 ff.; vgl. Ref. 3:1.

6 Treichler, R., vgl. Ref. 2:26.

7 Steiner, R. *The Wisdom of Man, of the Soul and of the Spirit* tr. S. and L. Lockwood. Lecture of 25.10.1909. New York: Anthroposophic Press 1971.

8 Steiner, R. *The Study of Man. General Education Course.* Tr. D. Harwood and H.Fox. Lectures of 25.8.1919. London, Rudolf Steiner Press 1975.

9 Strunk, P. in: *Lehrbuch der speziellen Kinder- und Jugendpsychiatrie*, S. 157 ff. Berlin/Heidelberg/New York 1971.

10 Sedlmayr. H. *Verlust der Mitte.* Salzburg 1955.

11 Lutz, I. *Kinderpsychiatrie*, S. 364 ff. Zuerich 1972. Zur Drogensucht vgl. ferner Kielholz, P., und Ladewig D. *Die Drogenabhaengigkeit des modernen Menschen.* Muenchen 1972.

12 Novalis. Fragment 623.

13 Steiner, R. *Spiritual Science and Medicine* Lecture of 9.4.1920. London: Rudolf Steiner Press 1948.

14 Suchantke, A. in: *Bewusstseinserweiterung durch Drogen?*, S. 32 ff. Stuttgart 1974.

15 Buehler, W. *Meditation als Erkenntnisweg. Bewusstseinserweiterung mit der Droge*, S. 30 ff. Stuttgart 1974.

16 Niederhaeuser, H.R. in: *Bewusstseinserweiterung durch Drogen?*, S. 27.

17 Buehler, W. Die Suche nach dem Bild. S. 29 ff. In: *Rauschgift.* Stuttgart 1980. Die Mitautoren: L.F.C. Mees und W. Schimpeler schildern andere Aspekte der Drogenproblematik. Vergleiche ferner Koob, O. *Droge und Suchtentstehung. Soziale Hygiene.* Bad Liebenzell 1981.

18 Steiner, R. *The Younger Generation.* Tr. R.M. Querido. Lecture of 8.10.1922. New York: Anthroposophic Press 1976.

19 Vgl. hierzu und zu den betreffenden Pflanzen Mees, L.F.C. *Rauschmittel - warum?* Stuttgart 1975.

20 Steiner, R. *Natur und Mensch in geisteswissenschaftlicher Betrachtung.* 8. Vortr. GA 352.

21 Heymann, K. in: *Bewusstseinserweiterung durch Drogen*, S. 17; vgl. Ref. 3:14.

22 Steiner, R. *Occult Science. An Outline.* Tr. G. and M. Adams. London: Rudolf Steiner Press 1969.

23 Steiner R. *The Education of the Child in the Light of Anthroposophy* Tr. M. and G. Adams, London: Rudolf Steiner Press 1975.

24 De Rudder, B. in: *Das Kind und die Zivilisation.* Koeln 1959.

25 Heymann, K. in: *Infantilismus*, S. 21; vgl. Ref. 3:1.

26 Baacke, H., Scheer, T. in: *Infantilismus*, S. 94; vgl. Ref. 3:1.

27 Steiner, R. 'The hidden depths of soul life'. Tr. A. Innes. Manuscript translation Z 303. Lecture of 23.11.1911. London: Rudolf Steiner House Library.

28 Ausfuehrlicher bei Treichler, R. in Husemann/Wolff. Vgl. Ref. 2:26.

29 Steiner, R. *Spiritual Science and Medicine* Lecture of 24.3.1920. London: Rudolf Steiner Press 1948. Grundlegendes zu diesem und anderen Problemen der anthroposophischen Medizin vgl. Sieweke, H. *Anthroposophische Medizin.* Dornach 1959.

30 Lievegoed, B. *Lebenskrisen-Lebenschancen*, S. 125 ff.; und Fintelmann, K.J.

Die Hibernia-Schule als Modell einer Gesamtschule. Stuttgart 1969.
31 Steiner, R. *Soul Economy and Waldorf Education* Lecture of 3.1.1922. London Rudolf Steiner Press 1986.
32 Vischer A.L. Vgl. Ref. 2:92.
33 Horney, K. *Der neurotische Mensch unserer Zeit*, S. 33 ff. Stuttgart 1951.
34 Koenig, K. Vgl. Ref. 2:19.
35 Zeylmans van Emmichhoven, F.W. *Gespraeche ueber die Hygiene der Seele*, S. 192. Arlesheim 1957.
36 Horney, K. *Der neurotische Mensch ...*, S. 175. Vgl. Ref. 3:33.
37 Hessenbruch, H. *Angst und Furcht*, S. 14. Ahrweiler 1952.
38 Ausfuehrlicher bei Treichler, R. in Husemann/Wolff. Vgl. Ref. 2:26.
39 Koenig, K. Vgl. Ref. 2:19.
40 Steiner, R. *Wonders of the World, Ordeals of the Soul, Revelations of the Spirit.* Tr. D. Lenn and O. Barfield. Lecture of 27.8.1911. London: Rudolf Steiner Press 1963.
41 V.d.Heide, P. Vgl. Ref. 2:66.
42 Freud, S. *Theoretische Schriften*, S. 363; vgl. Ref. 2:45.
43 Steiner, R. *Vom Lebenslauf des Menschen.* S. 17. Stuttgart 1980.
44 Steiner, R. *Metamorphoses of the Soul. Paths of Experience* Vol..2. Tr. C. Davy and C. v.Arnim. London: Rudolf Steiner Press 1983.
45 Horney, K. *Der neurotische Mensch ...*, S. 175; vgl. Ref. 3:33.
46 Steiner, R. *Metamorphoses of the Soul. Paths of Experience* Vol..2. Tr. C. Davy and C. v.Arnim. London: Rudolf Steiner Press 1983.
47 Frankl, V.E. *Anthroposophische Grundlagen der Psychotherapie*, S. 13 ff. Bern 1975.
48 Frankl, V.E. Die Heimholung der Psychotherapie in der Medizin. *Acta psychotherap.* 1962
49 Frankl, V.E. *Anthroposophische Grundlagen ...*, S. 256. Vgl. Ref. 3:47.
50 Frankl, V.E. ibido, S. 364 ff.
51 Steiner, R. *The Philosophy of Freedom.* Tr. M. Wilson. Ch. 4 and 5. London: Rudolf Steiner Press 1970.
52 Steiner, R. *Spiritual Science and Medicine* Lecture of 5.4.1920. London: Rudolf Steiner Press 1948.
53 Steiner, R. *The Wisdom of Man, of the Soul and of the Spirit* tr. S. and L. Lockwood. Lecture of 27.10.1909. New York: Anthroposophic Press 1971.
54 Steiner, R. *Psychoanalysis in the Light of Anthroposophy.* Tr. M. Laid-Brown. Lecture of 10.11.1917. New York: Anthroposophic Press 1946.
55 Steiner, R. *Metamorphoses of the Soul. Paths of Experience* Vol..2. Tr. C. Davy and C. v.Arnim. London: Rudolf Steiner Press 1983.
56 Steiner, R. *Occult Science. An Outline.* Tr. G. and M. Adams. London: Rudolf Steiner Press 1969.
57 Steiner, R. *Knowledge of the Higher Worlds.* Tr. G. Metaxa. London: Rudolf Steiner Press 1976.
58 Steiner, R. *Overcoming Nervousness.* Tr. R.Querido and F. Church. New York: Anthroposophic Press 1971.
59 Steiner, R. *The Philosophy of Freedom.* Tr. M. Wilson. Ch. 4 and 5. London: Rudolf Steiner Press 1970.

60 Steiner, R. *The Wisdom of Man, of the Soul and of the Spirit* tr. S. and L. Lockwood. Lecture of 26.10.1909. New York: Anthroposophic Press 1971.
61 King, M.L. Das Haus der Welt; zit. n. Nordmeyer, B.: *Sternkalender*. Dornach 1980.
62 Steiner, R. *The Younger Generation*. Tr. R.M. Querido. Lecture of 3.10.1922. New York: Anthroposophic Press 1976.
63 Glas, N. *Jugendzeit ...*, S. 39/40; vgl. Ref. 2:70.
64 Sheehy, G. Vgl. Ref. 1:19.
65 Steiner, R. *Menschliche und menschheitliche Entwicklungswahrheiten*. 1. Vortr. GA 176.
66 Unter Mitbenutzung eines Vortrags von N. Glas an der Tagung der Gesellschaft Anthroposophischer Aerzte, Teinach 1977.
67 Lusseyran, J. *Gegen die Verschmutzung des Ich*. Stuttgart 1972.
68 Sheehy, G. *In der Mitte des Lebens*, S. 303; vgl. Ref. 1:19.
69 Bodamer, J. *Der Mensch ohne Ich*, S. 105. Freiburg 1958.
70 Guardini, R. *Die Lebensalter*, S. 40; vgl. Ref. 1:14.
71 Lauer, H.E. *Vom richtigen Altwerden*, S. 44; vgl. Ref. 2:99.
72 Steiner, R. *Vom Lebenslauf des Menschen*. S. 126. Stuttgart 1980.
73 Zu den Phaenomenen und zur Haeufigkeit vgl. Richer, H./Beckmann, D. *Herzneurose*, S. 19ff. Stuttgart 1969.
74 Kleinsorge/Klumbies *Psychotherapie in Klinik und Praxis*, S. 10 ff. Muenchen/Berlin 1959.
75 Braeutigam, W., Christians, P. *Psychosomatische Medizin*, S. 105 ff. Stuttgart 1975.
76 Ibido S. 138 ff.
77 Steiner, R. *The Challenge of the Times*. Tr. O.D. Wannamaker. Lecture of 6.12.1918. New York: Anthroposophic Press 1987.
78 Steiner, R. *The Evolution of Consciousness*. Tr. V.E. Watkin. Lecture of 24.8.1923. London: Rudolf Steiner Press 1966.
79 Steiner, R. *The Challenge of the Times*. Tr. O.D. Wannamaker. Lecture of 7.12.1918. New York: Anthroposophic Press 1987.
80 Steiner, R. *Geisteswissenschaftliche Behandlung sozialer und paedagogischer Fragen*. 9. Vortr., GA 192.
81 Goethes Gespraeche mit Eckermann, aus dem Jahr 1828.
82 Steiner, R. *Geisteswissenschaftliche Behandlung sozialer und pädagogischer Fragen*. 11. Vortr., GA 192.
83 Mayer, K. Zur Psychopathologie vorzeitiger Versagenszustaende; in: *Zur Psychologie der Lebenskrisen*, S. 167. Frankfurt/M. 1962.
84 Schreiber, H. *Die Krise in der Mitte des Lebens*, S. 141. Muenchen 1977.
85 Lievegoed *Lebenskrisen ...*, S. 75; vgl. Ref. 1:15.
86 Buehler, C. *Die Psychologie im Leben unserer Zeit*, S. 294; vgl. Ref. 2:60.
87 Dubitscher, F. *Lebensschwierigkeiten und Selbsttoetung*, S. 12. Stuttgart 1971.
88 Sheehy, G. *In der Mitte des Lebens*, S. 267/268; vgl. Ref. 1:19.
89 Glas, N. *Jugendzeit ...*, S. 131; vgl. Ref. 2:70.
90 Steiner, R. *Friedrich Nietzsche*. Tr. M. Ingram de Ris. New Jersey: Rudolf Steiner Publications 1960.
91 Ausführlicher in: R. Treichler: Friedrich Hölderlin. Leben und Dichtung.

Krankheit und Schicksal. Stuttgart 1987.

92 Treichler, R. Friedrich Hoelderlin–Krankheit und Dichtung; zu dem Buch von P. Bertaux: Friedrich Hoelderlin; in: *Die Drei* 7/8/1979.

93 Husemann/Wolff; vgl. Ref. 2:3.

94 Steiner, R. *Occult Science. An Outline.* Tr. G. and M. Adams. Chapter on the stages in human life. London: Rudolf Steiner Press 1969.

95 Treichler, R. Vom Wesen der Neurasthenie; in: *Anthroposophie und Medizin*. Dornach 1963.

96 Steiner, R. *Knowledge of the Higher Worlds.* Tr. G. Metaxa. London: Rudolf Steiner Press 1976.

Chapter 4

1 Steiner, R. *Health Care as a Social Issue.* Lecture of 7.4.1920. Mercury 1984.

2 Steiner, R. 'Eight lectures to doctors'. Manuscript translation R 96. Lecture of 3.1.1924. London: Rudolf Steiner House Library.

3 Braeutigam, W. and Christians, P. *Psychosomatische Medizin*, S. 179; vgl. Ref. 3:75. Rattner, J. *Psychosomatische Medizin heute*, S. 81 ff. Zuerich/Stuttgart 1969.

4 Steiner, R., in Degenaar, A.G. Krankheitsfaelle, Manuskriptdruck fuer Aerzte, S. 75.

5 Steiner, R. *Physiologisch-Therapeutisches auf Grundlage der Geisteswissenschaft.* S. 236. GA 314.

6 Steiner, R. *Meditative Betrachtungen und Anleitungen zur Vertiefung der Heilkunst.* Osterkurs. 5. Vortr. GA 316.

7 Steiner, R. 'Eight lectures to doctors'. Manuscript translation R 96. Lecture of 9.1.1924. London: Rudolf Steiner Library.

8 Steiner, R. *Physiologisch-Therapeutisches auf Grundlage der Geisteswissenschaft.* S. 204. GA 314.

9 Steiner, R. *Die Ergänzung heutiger Wissenschaften durch Anthroposophie.* S. 290 ff. GA 73.

10 Steiner, R. *Soul Economy and Waldorf Education* Lecture of 31.12.1921. London Rudolf Steiner Press 1986.

11 E.g. in Steiner R. *Curative Education* Tr. M. Adams. Lecture of 25.6.1924 ('subtle defect in the liver'). London: Rudolf Steiner Press 1972

12 Steiner, R. *Rhythmen im Kosmos und im Menschenwesen.* Lecture of 28.6.1923. GA 350.

13 Steiner, R. *Spiritual Science and Medicine* Lecture of 9.4.1920. London: Rudolf Steiner Press 1948.

14 Steiner, R. Ibid. Lecture of 2.4.1920.

15 Steiner, R. *The Renewal of Education.* Answers to questions. Rudolf Steiner Schools Fellowship 1981.

16 See also Steiner, R., Wegman, I. *Fundamentals of Therapy.* Tr. E. Frommer and J. Josephson. London: Rudolf Steiner Press 1983.

17 Steiner, R. *Von Seelenrätseln* Ausg. 1917, Kap. IV, S. 235. GA 21.

18 Steiner, R. *Psychoanalysis in the Light of Anthroposophy*. Tr. M. Laird-Brown. Lecture of 2.7.1921. New York: Anthroposophic Press 1946.
19 Steiner, R. *Spiritual Science and Medicine* Lecture of 6.4.1920. London: Rudolf Steiner Press 1948.
20 Steiner, R. *Metamorphoses of the Soul. Paths of Experience* Vol..2. Tr. C. Davy and C. v.Arnim. London: Rudolf Steiner Press 1983.
21 Steiner, R. *Soul Economy and Waldorf Education* Lecture of 31.12.1921. London Rudolf Steiner Press 1986.
22 Treichler, R. Von einer Psychologie der Organe zu einer organischen Behandlung psychischer Stoerungen; in: *Anthroposophisch-medizinisches Jahrbuch*. Dornach 1952.
23 Steiner, R. 'Eight lectures to doctors'. Manuscript translation R 96. Lecture of 2.1.1924. London: Rudolf Steiner Library.
24 Jaspers, K. *Strindberg und van Gogh*. Muenchen 1926.
25 Steiner, R. *Manifestations of Karma*. Lecture of 20.5.1910. London: Rudolf Steiner Press 1976.
26 Das therapeutische Gespraech. Stuttgart 1980.
27 Weitere Ausfuehrungen in *Soziale Hygiene*, S. 13-46. Stuttgart 1973.

Chapter 5

1 Ausfuehrlicher bei Treichler, R. in Husemann/Wolff. Vgl. Ref. 2:26.
2 Kretschmer, E. *Medizinische Psychologie*, S. 204; vgl. Ref. 3:4.
3 Steiner, R. *Boundaries of Natural Science*. Tr. F. Amrine and K. Oberhuber. Lecture of 2.10.1920. New York: Anthroposophic Press 1983.
4 Nissen, G. in: *Lehrbuch der speziellen Kinder-und Jugendpsychiatrie*, S. 60 ff.; vgl. Ref. 3:9.
5 Braeutigam, W. u.a. *Psychosomatische Medizin*, S. 160; vgl. Ref. 3:75.
6 Steiner, R. Zur Therapie, in *Physiologisch-Therapeutisches*. S. 203. GA 314.
7 Steiner, R. *Spiritual Science and Medicine* Lecture of 31.3.1920. London: Rudolf Steiner Press 1948.
8 Bleuler, E. *Lehrbuch der Psychiatrie*, S. 527. Berlin 1975. Neubearbeitet von M. Bleuler.
9 Steiner, R. *Spiritual Science and Medicine* Lecture of 5.4.1920. London: Rudolf Steiner Press 1948.
10 Steiner R. *Curative Education* Tr. M. Adams. Lecture of 30.6.1924. London: Rudolf Steiner Press 1972
11 Nissen, G. in: *Lehrbuch der spez. Kinder- und Jugendpsychiatrie*, S. 71; vgl. Ref. 3:9.
12 Mueller-Wiedemann, H. *Mitte der Kindheit*, S. 20 ff.; vgl. Ref. 2:24.
13 Steiner R. *Curative Education* Tr. M. Adams. Lecture of 28 6.1924. London: Rudolf Steiner Press 1972
14 Steiner, R. Mental disorders viewed from the Standpoint of Spiritual Science. Tr. D. Osmond. *Anthroposophical Movement* 30:8,9.
15 Frankl, V.E. *Menschenbild der Seelenheilkunde*, S. 38/39. Stuttgart 1959.
16 Steiner, R., *Psychoanalysis in the Light of Anthroposophy*. See Ref. 4:18.
17 Steiner, R., 'Eight lectures to doctors', see Ref. 4:2.

18 Steiner, R., *Psychoanalysis in the Light of Anthroposophy*. See Ref. 3:54.
19 Steiner, R., Wegman, I. *Fundamentals of Therapy*. Tr. E. Frommer and J. Josephson. Chapter 2.London: Rudolf Steiner Press 1983.
20 Nissen, G. *Lehrbuch der spez. Kinder-und Jugendpsychiatrie*, S. 76; vgl. Ref. 3:9.
21 Kretschmer, E. *Hysterie, Reflex und Instinkt*, S. 34 ff. Stuttgart 1958.
22 Holtzapfel, W. *Seelenpflegebeduerftige Kinder*, S. 79 ff. Dornach 1976.
23 Steiner, R. *Spiritual Science and Medicine* Lecture of 22.3.1920. London: Rudolf Steiner Press 1948.
24 Steiner, R. *Physiologisch-Therapeutisches*. S.35. GA 314.
25 Steiner, R. 'The spiritual-scientific aspect of therapy'. 2nd Lecture of 18.4.1921. Manuscript translation. Association of Anthroposophical Doctors.
26 Weitere Ausfuehrungen und Begruendungen vgl. Treichler, R. *Vom Wesen der Hysterie*. Stuttgart 1964. Dort auch weitere Literatur.
27 Zum folgenden Kapitel vgl. Braeutigam, W./Christians, P. *Psychosomatische Medizin*, S. 245 ff. In seinem Aufsatz: Behandlung der Pubertaetsmagersucht (in *Beitraege* Heft 4/1980) geht J. Bockemuehl besonders auf entwicklungspsychologische Gesichtspunkte, darunter vor allem auf die gestoerte Atemreife ein. Auch die Therapie wird dort ausfuehrlicher dargestellt, als dies hier moeglich ist. Die umfassendste Darstellung gibt H. Koehler in: *Die stille Sehnsucht nach Heimkehr*. Stuttgart 1987.
28 Zu den Stoffwechselstoerungen bei Epilepsie, vgl. Selbach, H. Die zerebralen Anfallsleiden; in: *Handbuch der inneren Medizin*, Berlin 1953.
29 Steiner R. *Curative Education* Tr. M. Adams. Lecture of 27 6.1924. London: Rudolf Steiner Press 1972.
29a U. H. Peters: Die Epilepsie als psychisches Leiden. In: *Psychiatrie des 20. Jahrhunderts*. Weinheim u. Basel 1983.
30 Schulte, W. Epilepsie und ihre Randgebiete; in: *Klinik und Praxis*, S. 155. Muenchen 1964.
31 Holtzapfel, W. *Seelenpflegebeduerftige Kinder*, S. 61 ff.; vgl. Ref. 5:22.
32 Weitbrecht, H.J. *Psychiatrie im Umriss*, S. 282. Berlin 1973.
33 Kretschmer, E. *Koerperbau und Charakter*, 2. u. 13. Kap. Berlin 1944.
34 Steiner R. *Curative Education* Tr. M. Adams. Lecture of 26 6.1924. London: Rudolf Steiner Press 1972
35 Huber, G./Penin, H. in:*Psychiatrie der Gegenwart*, S. 644. Berlin 1973.
36 Weitere Ausfuehrungen und Begruendungen, vgl. Treichler R. *Vom Wesen der Epilepsie*. Neuauflage Stuttgart 1979. Dort auch weitere Literatur.
37 Bleuler, E. *Lehrbuch der Psychiatrie*, S. 409; vgl. Ref. 5:8.
38 Steiner, R. *Knowledge of the Higher Worlds*. Tr. G. Metaxa. London: Rudolf Steiner Press 1976.
39 Richter, D. *Schizophrenie, Somatische Gesichtspunkte*, S. 55. Stuttgart 1957.
39a Zu den verschiedenen Möglichkeiten, die Entstehung des Wahns zu verstehen vgl. P. Berner: Zur Psychologie des Wahns. In: *Psychologie des 20. Jahrhunderts. Psychiatrie I* S.515f. 1983.
40 Steiner, R. *Therapeutic Insights, Earthly and Cosmic Laws*. Lecture of 1.7.1921. Mercury 1984.
41 Steiner, R. *Metamorphoses of the Soul. Paths of Experience* Vol..2. Lectures

of 3.3. and 28.4. 1910. Tr. C. Davy and C. v.Arnim. London: Rudolf Steiner Press 1983.

42 Steiner, R. *The Stages of Higher Knowledge.* Tr. L. D. Monges and F. McKnight. New York: Anthroposophic Press 1974.

43 Steiner, R. 'The spiritual-scientific aspect of therapy'. Lecture of 16.4.1921. Manuscript translation. Association of Anthroposophical Doctors.

44 Steiner, R. *Spiritual Science and Medicine* Lecture of 24.3.1920. London: Rudolf Steiner Press 1948.

45 Zit. n. *Fortschritte der Neurologie und Psychiatrie,* S. 475/9, 1962.

46 Bleuler, M. *Die schizophrenen Geistesstoerungen,* S. 532. Berlin 1972.

47 Bleuler, E. *Lehrbuch der Psychiatrie,* S. 447; vgl. Ref. 5:8.

48 Kretschmer, E. *Koerperbau und Charakter,* S. 332 ff.; vgl. Ref. 5:33.

49 Steiner, R. *Spiritual Science and Medicine* Lecture of 5.4.1920. London: Rudolf Steiner Press 1948.

50 Bleuler, E. *Lehrbuch der Psychiatrie,* S. 326; vgl. Ref. 5:8.

51 Steiner, R. Mental disorders viewed from the Standpoint of Spiritual Science. Tr. D. Osmond. *Anthroposophical Movement* 30:8,9.

52 Steiner, R. Man in the cosmos as a being of thought and will. Tr. M. Cotterell. Manuscript translation Z 325. London: Rudolf Steiner House Library.

53 Steiner, R. *Rhythmen im Kosmos und im Menschenwesen.* Lecture of 28.6.1923. GA 350.

54 Laing, R. *Phaenomenologie der Erfahrung,* S. 122 ff. Frankfurt/M. 1975.

55 Poppelbaum, H. *Im Kampf um ein neues Bewusstsein,* S. 79. Freiburg i. Br. 1948.

56 Steiner, R. *Knowledge of the Higher Worlds.* Tr. G. Metaxa. London: Rudolf Steiner Press 1976.

57 Husemann, F. Psychiatrische Fragen vom Gesichtspunkt der Anthroposophie; in: *Kunst und Wissenschaft,* S. 307 ff. Stuttgart 1922.

58 Weitere Ausfuehrungen und Literatur zur Schizophrenie in: Treichler, R. *Der schizopohrene Prozess.* Neuaufl. Stuttgart 1981. Die letzten Abschnitte sind teilweise jenem Buch entnommen.

59 Steiner, R. *Manifestations of Karma.* Lecture of 25.5.1910. London: Rudolf Steiner Press 1976.

60 Schulte, W. *Ueber das Wesen melancholischen Erlebens und die Moeglichkeiten der Beeinflussung,* S. 83 ff. Stuttgart 1965.

61 Zur Phaenomenologie und zum Wesen der Manie vgl. Blankenburg, W. Manie; im *Almanach f. Neurologie u. Psychiatrie* 1961. Zur Melancholie, insbesondere zum schwernehmenden Charakter, vgl. Tellenbach, H. *Melancholie.* Berlin 1976.

62 Kleinsorge, H./Klumbies, W. *Psychotherapie in Klinik und Praxis,* S. 108; vgl. Ref. 3:74.

63 Wolff, O. in Husemann/Wolff *Das Bild des Menschen ...,* II, S. 489 ff.; vgl. Ref. 2:3.

64 Steiner, R. *Spiritual Science and Medicine* Lecture of 9.4.1920. London: Rudolf Steiner Press 1948.

65 Wachsmuth, G. *Erde und Mensch,* Kap. VIII. Konstanz 1952. Zum

Verhalten der Wesensglieder vgl. auch: Treichler, R. Schlafen und Wachen als Atmung des Ich. Vgl. Ref. 2:65.
66 Bleuler, E. *Lehrbuch der Psychiatrie*, S. 471. Berlin 1975.
67 *Aerztliche Praxis*, 28.12.1976.
68 Treichler, R. Vgl. ref. 4:22.
69 Z. B. con I. H. Schulz im *Handbuch der Neurosenlehre und Psychotherapie*: Der medizinisch-psychologische Gesichtspunkt. S. 247. München–Berlin 1959.

Chapter 6

1 Vgl. Ref. 2:65
2 Steiner, R. *Das Hereinwirken geistiger Wesenheiten in den Menschen.* 11. Vortr. (11.6.1908). GA 102.
3 Beuys, J. *Jeder Mensch sein Kuenstler.* Frankfurt/M. 1975.
4 Hauschka, M. *Zur kuenstlerischen Therapie,* Boll 1971.
5 Ausfuehrlicher bei Treichler, R. in Husemann/Wolff. Vgl. Ref. 2:26.
6 Vgl. auch zum folgenden: Schwebsch, E. *Zur aesthetischen Erziehung des Menschen.* Stuttgart 1954.
7 Hauschka, M. *Wesen und Aufgaben der Maltherapie; zur kuenstlerischen Therapie,* Bd II, Boll 1978. Grundlegendes ferner bei v.d.Heide, P. *Zur kuenstlerischen Therapie,* Bd III, Boll 1978.
8 Steiner, R. *Kunst und Kunsterkenntnis.* S. 107 ff. GA 271. This lecture cycle contains statements on the physical connections of the different arts. Two of the lectures have been translated into English: 1) Lecture of 28.10.1909 as *Nature and Origin of the Arts,* London: Anthroposophical Society, no date; and 'The being of the arts', tr. A. Bittleston, *Golden Blade* 1979. 2) 'The physical-superphysical: Its relation in art', tr. V.E. Watkin, manuscript translation Z 385, London: Rudolf Steiner House Library.
9 Zu einer anthroposophisch orientierten Musiktherapie vgl. z.B. Oberkogler, F. *Heilende Kraefte der Musik.* Wien 1978.
10 Steiner, R. *Die Entstehung und Entwicklung der Eurythmie.* S. 111. Dornach 1965.
11 Clauser, G. (Die vorgeburtliche Entstehung der Sprache als anthropologisches Phaenomen. Stuttgart 1971.
12 Steiner, R. *Speech and Drama.* Tr. M. Adams. London: Anthroposophical Publishing Co. 1960. Koenig/v.Arnim/Herberg Sprachverstaendnis und Sprachbehandlung; in *Heilpaedagogik aus anthroposophischer Menschenkunde, Bd 4. Stuttgart 1978.–Lorenz-Poschmann, A. Therapie durch Sprachgestaltung.* Dornach 1981.
13 Steiner, R. *Curative Eurythmy.* Tr. K. Krohn and A. Degenaar. London: Rudolf Steiner Press 1983. Kirchner-Bockholt, M. *Grundelemente der Heileurythmie.* Dornach 1972.–Treichler, R. Heileurythmie in der Psychiatrie; in: *Arzt und Heileurythmie.* Dornach 1972.
14 Bothmer, Fr. Graf v. *Gymnastische Erziehung* (1959). Stuttgart 1981.

Chapter 7

1 Steiner, R. *The World of the Senses and the World of the Spirit.* Lecture of 27.12.1912. London: Rudolf Steiner Publishing Co. 1947.

2 Steiner, R. *Soul Economy and Waldorf Education* Lecture of 26.12.1921. London Rudolf Steiner Press 1986.

3 Steiner, R. *Occult Science. An Outline.* Tr. G. and M. Adams. London: Rudolf Steiner Press 1969.

4 Steiner, R. 'Swiss teachers' Course, report by Albert Steffen. Manuscript translation R 34. Lecture of 16.4.1923. London: Rudolf Steiner House Library.

5 Steiner, R. *Geisteswissenschaft und die Lebensforderungen der Gegenwart.* S. 40 ff. Dornach 1950.

6 Steiner, R. *Das lebendige Wesen der Anthroposophie und seine Pflege.* An die Mitglieder VII. GA 37.

7 Steiner, R. *The Evolution of Consciousness.* Tr. V.E. Watkin. Lecture of 21.8.1923. London: Rudolf Steiner Press 1966.

8 Kuehlewind, G. *Die Wahrheit tun,* S. 68 ff. Stuttgart 1978.

9 Steiner, R. *Wie erwirbt man sich Verständnis für die geistige Welt?.* S. 67. GA 154.

10 Steiner, R. *Menschliche und menschheitliche Entwicklungswahrheiten.* 4. Vortr. GA 176.

11 Steiner, R. *Das lebendige Wesen der Anthroposophie und seine Pflege.* 1. Leitsatz. GA 37.

11a Steiner, R., *Knowledge of the Higher Worlds.* Tr. G. Metaxa. Rudolf Steiner Press 1976.

12 Steiner, R. *Anthroposophical Approach to Medicine.* Tr. C. Davy. Lecture of 26.10.1922. London: Anthroposophical Publishing Co., 1951.

13 Steiner, R. *The Philosophy of Freedom.* Tr. M. Wilson. Ch. 4 and 5. London: Rudolf Steiner Press 1970.

14 Witzenmann, H. *Intuition und Beobachtung,* S. 25. Stuttgart 1978.

15 Steiner, R. *Die Schwelle der geistigen Welt.* Ausg. 1935, S. 23/24. GA 17.

16 Schiller, P.E. *Der anthroposophische Schulungsweg,* S. 68 und 113. Dornach 1979.

17 Steiner, R. *The Evolution of Consciousness.* Tr. V.E. Watkin. Lecture of 20.8.1923. London: Rudolf Steiner Press 1966.

18 Steiner, R. 'The hidden depths of soul life'. Tr. A. Innes. Manuscript translation Z 303. Lecture of 23.11.1911. London: Rudolf Steiner House Library.

19 Steiner, R. *Die Erkenntnis des Uebersinnlichen in unserer Zeit. Gesundheit und Krankheit im Seelenleben.* Vortr. 14.2.1907. GA 55.

20 Steiner, R. *The Evolution of Consciousness.* Tr. V.E. Watkin. Lecture of 19.8.1923. London: Rudolf Steiner Press 1966.

21 Steiner, R. *The Challenge of the Times.* Tr. O.D. Wannamaker. Lecture of 6.12.1918. New York: Anthroposophic Press 1987.

22 Steiner, R. *Towards Social Renewal.* Tr. F.T. Smith. London: Rudolf Steiner Press 1977.

23 Steiner, R. *The Challenge of the Times.* Tr. O.D. Wannamaker. Lecture of 7.12.1918. New York: Anthroposophic Press 1987.

24 Steiner, R. *The Effects of Spiritual Development.* Tr. A.H. Parker. Lecture of 29.3.1913. London: Rudolf Steiner Press 1978.

25 Steiner, R. *Occult Science. An Outline.* Tr. G. and M. Adams. London: Rudolf Steiner Press 1969.

26 Steiner, R. *The Study of Man. General Education Course.* Tr. D. Harwood and H.Fox. Lecture of 27.8.1919. London, Rudolf Steiner Press 1975.

27 Schiller, P.E. *Der anthroposophische Schulungsweg,* S. 128; vgl. Ref. 7:16.

28 Steiner, R. *Was wollte das Goetheanum und was soll die Anthroposophie?.* S.230 ff. GA 84.

29 Steiner, R. *Occult Science. An Outline.* Tr. G. and M. Adams. London: Rudolf Steiner Press 1969.

30 Steiner, R. *Theosophy* Tr. A.P. Shepherd. London: Rudolf Steiner Press 1973.

31 Steiner, R. *The Stages of Higher Knowledge.* Tr. L.D. Monges and F. McKnight. New York: Anthroposophic Press 1974.

32

33 Steiner, R. *Die Erkenntnis des Uebersinnlichen in unserer Zeit.* Der Lebenslauf des Menschen vom geisteswissenschaftlichen Gesichtspunkt. GA 55.

34 Steiner, R. *Boundaries of Natural Science.* Tr. F. Amrine and K. Oberhuber. 1st lecture of 2.10.1920. New York: Anthroposophic Press 1983.

35 Steiner, R. *Boundaries of Natural Science.* Tr. F. Amrine and K. Oberhuber. Lecture of 1.10.1920. New York: Anthroposophic Press 1983.

36 Novalis. Fragment Nr. 1625.

37 Steiner, R. *True and False Paths in Spiritual Investigation.* Tr. A.H. Parker. Lecture of 18.8.1924. London: Rudolf Steiner Press 1969. R. Treichler: Seelische Entwicklung und geistige Schulung. In: Jörgen Smit u.a.: *Freiheit erüben.* Stuttgart 1988.

38 Steiner, R. *Occult Science. An Outline.* Tr. G. and M. Adams. London: Rudolf Steiner Press 1969.–Treichler, R. Lebenslauf, Seelenentwicklung und seelische Stoerungen III. *Das Goetheanum,* 5.12.1976.

About the Author:
Rudolf Treichler

Rudolf Treichler was born on 10 March 1909, son of the Waldorf School teacher Rudolf Treichler, PhD. After attending the Stuttgart Waldorf School from 1920 to 1928, he studied medicine and wrote his dissertation on *Friedrich Hoelderlin's mental condition in relation to his poetic works*. Specialist training in psychiatry and neurology. Eight years at the Stuttgart Hospital for Nervous Diseases. Fourteen years of private practice as an anthroposophical specialist for nervous diseases in Stuttgart. Director of annual Week of Psychiatric Studies at the Goetheanum in Dornach. 1959-1974 Joint Medical Director of Friedrich Husemann Clinic with Dr Priever. Since then worked at the Clinic in consultant capacity. Courses in psychiatry and lectures at home and abroad. Publications in the field of psychiatry and neurology, contributions to symposia, essays in journals, volumes of poetry.

About the Translators

Anna Meuss was born in Germany and came to the UK at the age of 19. She was asked to translate for Dr Karl Koenig and worked with him for a short time at Camphill, and then for many years with Dr Ralph Twentyman.

Now works as a freelance translator, editor and writer, Visiting Lecturer in Scientific German at Kingston Polytechnic, and examiner for the Institute of Linguists.

Fellow of the Institute of Linguists, Member of the Society of Authors and the Institute of Translation and Interpreting. Member of Translators and Editors in Anthroposophy and co-editor of its journal *Word for Word*.

Johanna Collis was born in 1937. She attended Michael Hall Rudolf Steiner School in England and trained as a translator at Sprachwissenschaftliches Institut für das Dolmetscher-und Übersetzerwesen, Berlin, and Sprachen-und Dolmetscher Institut, Munich, gaining her translator's diploma in 1961. This was followed by three years as a translator on the staff of the Deutsche Stiftung für Entwicklungsländer in Berlin. Since March 1987 freelance translator in the field of anthroposophy.

Member of the Institute of Linguists and of Translators and Editors in Anthroposophy.

Translator's Note on Terminology

Note on the terminology used in this translation
  Below a list is given of some of the terms used in this translation
and their equivalents in anthroposophical usage.

Term used in this work	*Anthroposophical usage*
(etheric) creative powers	formative forces
awareness soul	consciousness soul
essential, or constituent, principles	members of being
heart and mind, or (if the context suggested a definite will element) heart and soul	Gemüt

'Egoity' and 'egohood' have been used in the differentiated fashion
suggested by the *Shorter Oxford English Dictionary*, where the entry
on 'ego' includes 'hence "egohood", 'individuality', and 'egoity' has
its own entry: 'selfhood; that which forms the essence of the
individual.'

Books in Print:
Social Ecology Series

**BATTLE FOR THE SOUL The working together of three great
leaders of humanity**
Bernard Lievegoed.

Bernard Lievegoed M.D. was a distinguished physician, psychiatrist,
organisational development consultant and educationalist. *The Battle of
the Soul* complements his more autobiographical, *The Eye of the Needle*
(Hawthorn Press, 1993).
135 x 238mm; 144pp; ISBN 1 869 890 64 7

DEVELOPING COMMUNITIES
Bernard Lievegoed.

Forming Curative Communities addresses the development of all forms of
communities and groups that come into existence with a spiritual/cultur-
al aim. *Concerning Organisations of the Spiritual Life* offers further insights
and direction about community living. Lievegoed suggests approaches to
working together and staff training methods that will truly help people to
work together.
216 x 138mm; 218pp; ISBN 1 869 890 30 2

**EYE OF THE NEEDLE   His life and working encounter with
anthroposophy**
Bernard Lievegoed.

A man of wide ranging interests, Lievegoed combined his profound
inner, spiritual research with his pioneering social, medical, educational
and management work to produce a number of fascinating books. *The
Eye of the Needle* illustrates the dynamics between the inner and outer
worlds - and of Lievegoed's ability to work with these dynamics.
216 x 138mm; 103pp; paperback; ISBN 1 869 890 50 7

HOPE, EVOLUTION AND CHANGE
John Davy.

The twenty-seven articles in this book reflect the author's work as a
scientist, journalist and lecturer: articles on evolutionary questions, language, education, science, ecology, life after life and contemporary
thinkers such as Schumacher and Elizabeth Kübler-Ross.
210 x 135mm; 274pp; paperback; ISBN 1 869 890 27 2

HOW TO TRANSFORM THINKING, FEELING AND WILLING
Jorgen Smit.

This book aims to enable readers to follow a meditative path leading to deepening insight and awareness of themselves and the world around them. Practical exercises for illuminating and strengthening thinking are described, for developing inspiration and intuition and also for exploring the qualities of composure, reverence, openmindedness and wonder.
210 x 135mm; 64pp; paperback; ISBN 1 869 890 17 5

IN PLACE OF THE SELF How drugs work
Ron Dunselman.

Why are heroin, alcohol, hashish, ecstasy, LSD and tobacco attractive substances for so many people? Why are unusual, visionary and 'high' experiences so important to users? How can we understand such experiences? These and others questions about drugs and drug use are answered comprehensively in this remarkable book by Ron Dunselman
216 x 138mm; 304pp; hardback; ISBN 1 869 890 72 8

MAN ON THE THRESHOLD
Bernard Lievegoed.

Concerned with inner training and development, *Man on the Threshold* takes an anthroposophical approach to its theme. Lievegoed, the distinguished physician, educator and industrial psychologist, is aware that humanity is crossing a major threshold and a redefinition of boundaries is needed.   210 x 135mm; 224pp; paperback; ISBN 0 950 706 26 4

MONEY FOR A BETTER WORLD
Rudolf Mees.

This slim volume re-works our attitudes towards handling money and presents finance on a human scale. It discusses alternative ways of looking at money in our modern world and realistic methods of approaching borrowing, saving and lending.
216 x 138mm; 64pp; paperback; ISBN 1 869 890 26 4

MORE PRECIOUS THAN LIGHT
How dialogue can transform relationships and build commuity
Margreet van den Brink.

Profound changes are taking place as people awaken to the experience of the Christ in themselves. The author is a social consultant and counsellor and offers helpful insights into building relationships. She shows how true encounter can be fostered.
216 x 138mm; 160pp; colour cover; ISBN 1 869 890 83 3

RUDOLF STEINER An Introduction
Rudi Lissau.

This portrait of Steiner's life and work aims to point out the relevance of his activities to contemporary social and human concerns. Under discussion are Steiner's philosophy; his view of the universe, earth and the human being; Christ and human destiny; the meditative path; education and social development and approaches and obstacles to his work.
210 x 135mm; 192pp; ISBN 1 869 890 06 8

SEVEN SOUL TYPES
Max Stibbe. Translated by Jakob Cornelis.

The educationalist, Max Stibbe, describes the seven soul types of man, indicating the most significant inner and outer characteristics of each type. Recognition of soul types can be invaluable in communicating with others in social, educational or therapeutic situations and the author's insights are both fascinating and instructive.
216 x 138mm; 128pp; ISBN 1 869 890 44 2

THE ENTERPRISE OF THE FUTURE
Moral intuition in leadership and organisational development
Friedrich Glasl

Friedrich Glasl describes the future of the modern organisation as a unique challenge for personal development. Every organisation, whether a business, a school, a hospital or a voluntary organisation, will have to develop closer relationships with the key stakeholders in its environment - its suppliers, customers, investors and local communities. Our consciousness as managers needs to expand beyond the boundaries of the organisation to work associatively with the community of enterprises with whom we 'share a destiny'.
216 x 138mm; 160pp; ISBN 1 869 890 79 5

TWELVE SENSES
Albert Soesman.

The senses nourish our experience and act as windows on the world. But our stimulation may undermine healthy sense experiences. The author provides a lively look at the senses, not merely the normal five senses, but twelve: touch, life, self-movement, balance, smell, taste, vision, temperature, hearing, language, the conceptual and the ego senses.
210 x 135mm; 176pp; ISBN 1 869 890 22 1

Lifeways/Parenting Series

ALL YEAR ROUND
Ann Druitt, Christine Fynes-Clinton, Marije Rowling

Brimming with seasonal stories, activities, crafts, poems and recipes, this book offers a truly inspirational guide to celebrating festivals throughout the seasons.

200 x 250mm; 288pp; colour cover; fully illustrated; ISBN 1 869 890 47 7

BETWEEN FORM AND FREEDOM A practical guide to the teenage years
Betty Staley

Betty Staley offers a wealth of insights about teenagers, providing a compassionate, intelligent and intuitive look into the minds of children and adolescents. Issues concerning stress, depression, drug and alcohol abuse and eating disorders are included.

210 x 135mm; 288pp; illustrated; ISBN 1 869 890 08 6

CHILD'S PLAY 3 Games for life for children and teenagers
Wil van Haren and Rudolf Kischnick.
Translated by Plym Peters and Tony Langham

A tried and tested games book consisting of numerous ideas for running races, duels, wrestling matches, activity and ball games and games of skill and agility. It's clear lay-out, detailed explanations and diagrams and its indexing of games by age suitability and title makes *Child's Play 3* an invaluable and enjoyable resource book for parents, teachers and play leaders.

145 x 215mm; 96pp; paperback; colour cover; ISBN 1 869 890 63 9

CHILD'S PLAY 1 AND 2 Games for children and teenagers
Wil van Haren and Rudolf Kischnick

Child's Play 1 and 2 follows up *Child's Play 3* as a games book particularly for younger children, suitable for nursery, kindergarten and junior schools. Parents, teachers and play leaders will find this guide to be a handy resource for schools, camps, parties, family occasions and gatherings.

215 x 145mm; 192pp; colour cover; paperback; ISBN 1 869 890 77 9

FESTIVALS TOGETHER A guide to multi-cultural celebration
Sue Fitzjohn, Minda Weston, Judy Large

This is a resource guide for celebration, and for observing special days according to traditions based on many cultures. It brings together the experience, sharing and activities of individuals from multi-faith communities all over the world - Buddhist, Christian, Hindu, Jewish, Muslim and Sikh.

200 x 250mm; 224pp; colour cover; fully illustrated; ISBN 1 869 890 46 9

GAMES CHILDREN PLAY
Kim Brooking Payne

Following on from *Looking Forward, Child's Play 3* and *Child's Play 1 and 2,* Kim Payne's *Games Children Play* offers an accessible guide to games with children ages 3 upwards. These games are all tried and tested with children, and are the basis for the author's extensive teacher training work.

215 x 145mm; 256pp; paperback; illustrations; ISBN 1 869 890 78 7

INCARNATING CHILD
Joan Salter.

'Our birth is but a sleep and a forgetting'; Joan Salter picks up on Wordsworth's theme and follows the soul life of tiny babies into childhood and adolescence. A specialist in maternal and child care, she addresses physical, spiritual and psychological development as well as environmental factors. This book will be particularly valuable for those embarking on parenthood for the first time.

210 x 135mm; 224pp; illustrations and photographs; ISBN 1 869 890 04 3

LIFEWAYS Working with family questions
Gudrun Davy and Bons Voors.

Lifeways is about children, about family life and about being a parent. But above all it is about freedom, and how the tension between family life and personal fulfilment can be resolved.

150 x 210mm; 328pp; ISBN 0 950 706 24 8

PARENTING FOR A HEALTHY FUTURE
Dotty T. Coplen

Here is a commonsense approach to the challenging art of parenting; an offer of genuine support and guidance to encourage parents to believe in themselves and their children. Dotty Coplen helps parents gain a deeper under-standing of parenting children from both a practical and holistic, spiritual perspective.

216 x 138mm; 126pp;   ISBN 1 869 890 53 1

VOYAGE THROUGH CHILDHOOD INTO THE ADULT WORLD A guide to child development
Eva A. Frommer

Human beings have a long infancy during which they are dependent upon others for the means of life and growth - such a book on child development is therefore vital. A deep concern for the uniqueness of each individual child permeates this book, while offering practical solutions to the challenges of raising a child at each stage of his or her development.

216 x 138mm; 152pp; paperback;
Colour and black & white photographs;   ISBN 1 869 890 59 0

Ordering

If you have difficulties ordering from a bookshop you can order direct from:

Hawthorn Press,
Hawthorn House,
1 Lansdown Lane,
Lansdown,
Stroud,
Glos., UK,
GL5 1BJ

Tel: (01453) 757040
Fax: (01453) 751138

INDEX